CU00751080

Energy Crisis:

Solution From Space

By Ralph Nansen

All rights reserved under article two of the Berne Copyright Convention (1971).
We acknowledge the financial support of the Government of Canada through the Book Publishing Industry Development Program for our publishing activities.

Published by Apogee Books an imprint of Collector's Guide Publishing Inc., Box 62034, Burlington, Ontario, Canada, L7R 4K2, http://www.cgpublishing.com

Printed and bound in Canada

Energy Crisis: Solution From Space by Ralph Nansen
ISBN 9781-926592-06-0 - ISSN 1496-6921

©2009 Apogee Books/Ralph Nansen

Cover: Robert Godwin

Energy Crisis:

Solution From Space

By Ralph Nansen

An Apogee Books Publication

Table of Contents

Acknowledgements

It has been 36 years since I first became aware of the concept of solar power satellites. For nearly half my life they have been the focus of my life. I could not have sustained the effort without the support and encouragement of my wife Phyllis. In addition she has been my chief editor for this book, fixing my poor grammar, putting me back on course when I would stray off or become redundant.

I want to thank Rob Godwin, President of CG Publishing for recognizing the importance of spreading the word about solar power satellites and agreeing to publish this book.

I want to acknowledge and thank the Aerospace Technology Working Group and particularly Feng Hsu, the primary author of the paper that is shown in its entirety in Chapter 10. This paper lays out a clear plan on what needs to be done to bring the United States back on course in the development and exploration of our space frontier.

Over the years since I wrote SUN POWER: The Global Solution for the Coming Energy Crisis, there have been many that have encouraged me to keep pressing on and urging me to write another book. One individual stands out. He is Neil Hanson who manages the Lopez Island Recycle and Transfer Station. Every time I drop off our trash and recycle material he has something to tell me or a question about how things are progressing in getting space solar power going. He is frustrated with the lack of knowledge in the media concerning energy from space. Thank you Neil for your support and encouragement. And I want thank all the others that keep helping push on to bring energy from space to the attention those that do not know there is a solution to our energy problems.

The sun, our original source of energy. It grew the trees our ancestors burned for light and warmth. It energized the life forms that became our coal and oil and natural gas. It warms the earth and evaporates the water to bring us rain and flowing rivers. It makes the wind blow. Mankind has used all of these energy forms, and now we need more. Now is the time to directly tap the sun's energy to power our future. I will show you the way.

1   Our Troubled World

Our world is in trouble. The specter of an economic collapse that could make the great depression of the 1930's look like a mild recession in comparison is just waiting to happen. All the elements are in place. It started with the implosion of the mortgage market as the giant banking and financial institutions tumble into bankruptcy and Fannie Mae and Freddie Mac were taken over by the government to bail them out from total collapse. It was helped along by the dramatic rise in the cost of oil as it soared to $148 a barrel and everyone in the world was impacted. As the situation evolved the world markets went into freefall and unemployment swept across the nation and on around the world. As the world slipped into recession our government desperately moved to provide a $700 billion bailout package. Then after the installation of the new administration an additional $787 billion stimulus package was passed to try and stop the fall.

The economy is not our only problem. We face the issue of our depleting the source of energy that sustains our modern world. This is compounded by the fact that it is the burning of our fossil fuel energy sources that have powered our civilization, are also in the process of polluting our atmosphere and causing global warming. This is leading to violent weather patterns and hurricanes that are devastating our people. The arctic and Antarctic ice is melting away and the danger of rising sea levels threaten all the earth's low lying lands and our sea ports. All glaciers on earth are receding and the water supply for 40% of the world's population is threatened. The United States is running a huge deficit. We are spending $700 billion on annual oil imports and at the same time borrowing enormous amounts from China to support our economy.

The Iraq war struggles on after years of occupation. The conflict in Afghanistan is escalating as the Taliban gains strength. The unknown of what Iran might do in the Middle East hangs over our heads. The danger of further wars over Middle East oil is a dark shadow over the future as nations become desperate for energy.

This is the future that stands before us if we continue on our current course. This book is about an ultimate solution to our problems, but let me start with a story of my early life that will give you an idea of what shaped my life and gives us a correlation to our future today.

● ● ● ● ● ● ● ● ● ● ● ●

I was born in the early years of the great depression in a little wheat farming town in the eastern part of the state of Washington. The first memory I have as a child was the afternoon my father drove up to our house in a shiny new 1935 maroon colored Chrysler. I thought it was the prettiest car I had ever seen. It was the middle of the depression and there weren't many new cars around. But our town was not typical for that time, because it was bustling with activity. We were only 18 miles from the construction site of Grand Coulee Dam and many of the workers were living in temporary housing or renting rooms in the homes of the town's people. The chief engineer of the dam lived at our house and rented one of our upstairs bedrooms.

Every Sunday afternoon the family would pile into the Chrysler and off we would go on a drive down to Grand Coulee to see the progress on the dam. I use the term "drive down" because that's the term everyone used. We did in fact have to drive down a winding switch-back highway that made its way down from the wheat land into the coulee where the dam was being built. The dam construction was like a great magnet drawing out-of-work and destitute people searching for work. They came in droves, many more than it was possible to employ on the site. Our town like many others had a "bum camp" down by the railroad tracks where they congregated. There was never a day that went by that there wasn't a knock on the door of our home with a worn, bedraggled man, standing there asking if we could give him some food or money. My mother never sent them away without giving them a meal. We had a big garden, chickens, and a cow. We raised enough food to supply nearly all of our needs. In addition my uncle was a farmer and he supplied us with a steer and a hog each fall. My mother canned the beef and I helped my dad set up and cure the hams and bacon. My mother made head cheese out of the hogs head and I thought that was the best stuff I had ever eaten.

My father and his partner owned the small bank in the town. It had been one of three banks in town when the depression started, but the other two had failed early in the depression. I can remember my father telling how their bank would have failed also if Roosevelt hadn't ordered the closure of all banks when he did. It gave my father time to restructure and stay afloat. Years later I learned how the bank had kept farmers going even when they could not pay back the loans they had with the bank. My father said it would not have helped to foreclose on a farmer's land as there was nobody with any money that could buy it. So the bank kept giving the farmers enough money to keep them going. After it was all over, they were able to keep their farms and only one farmer failed to repay all he owed to the bank.

While all this was going on Grand Coulee Dam was growing out of the river bed and crawling up the rock walls of the coulee. My father and his partner had been two of the advocates that had been working for years to have the dam built. In fact the land it was being built on had belonged to my dad's partner. He was paid about a hundred dollars for 160 acres by the government. The final impetus for building the dam came because of the need to create jobs during the depression.

This was a very exciting time for a young boy like myself. I was seeing the ravages of the depression and the grandeur of the largest power plant in the world being built at the same time. All of the material needed to build the dam, except for the sand and gravel used to make the cement, came through our town on the railroad. The tracks were only about four blocks from our house and the freight trains would come rumbling through at all hours of the day and night. I used to put pennies on the track to see what they would look like after a train passed over them and flattened them to a paper-thin sheet of copper. I will never forget the day when the first drive shafts for the generators came through town. I had never seen anything so big. Each shaft was solid steel and so big that it took two flat cars with pivots so that one end of the shaft rested on one car and the other on the second car.

One other part of growing up in a small eastern Washington town that ultimately influenced my life were the crystal clear nights with the moon and stars filling the sky with brilliant light, without the light pollution of a city to detract from them. I became very intrigued with the moon. I guess I didn't realize how much it had affected me until one night I was visiting a high school friend. We were outside admiring a full moon when his father walked up and asked what we were doing. Without even thinking I blurted out "we're looking at the moon and I'm going to go there someday." My friend's father looked at me and said, "You're full of hooey," and walked away. My off-hand remark and the response it received crystallized the direction I set for my life. The first step was to become an engineer. The second was to get into a company that built rockets. It was only 11 years after that off-hand remark until I was sitting at a drafting table in the Boeing plant drawing the configuration for the first stage of the rocket that sent the men to the moon. I didn't make it there myself, but I helped those that did.

The experience of growing up during the depression and seeing how bad it was for people and also seeing the magnificence that people can accomplish has probably molded my life more than I have ever realized. This book addresses the kind of problems we face today that can bring the tragedy of the great depression back on us. It will also show you the potential of building a magnificent new energy system that can lift us out of a threatened world to one of hope and lasting prosperity.

• • • • • • • • • • • • •

The first time that we as a people in the United States really thought about the possibility that abundant oil might not always be available to us was during the 1973-74 OPEC oil embargo, Saudi Arabia led the other Arab nations in OPEC to cut off the supply of oil to the United States and other Western nations in retaliation for our support of Israel. We suffered with long lines at the gas stations, escalating gas prices, reduced speed limits, lower settings on our thermostats, and the need to convert to more efficient automobiles. The government established gas mileage requirements for new automobiles and because of the high gas prices we experienced a significant economic impact. Japan's automotive industry which was already focused on small high mileage cars began the eclipse of the US auto industry. Other United States industries, such as steel reeled under the cost impact and never fully recovered. We had our first big scare, but our memory was short.

As a result of the oil embargo we were forced to make the changes in our energy use that did significantly reduce our use of energy to compensate for the shortages. This along with the easing of the embargo brought the prices down. After a time our consumption started to rise again.

We had another jolt in gas prices in 1990 at the time of the first Gulf War, when we went to war over oil. When it was over we once again dropped back into our routine use of oil like there was no end to it. As we moved into the decade of the 90's and on into the 21st century, we relentlessly pursued our love of big vehicles. The in-thing was big pickup trucks and then SUVs. They started with reasonable sizes but then grew and grew. Moms had to have one to take the kids to school and go to the mall. The engines were big and their weight was monstrous and fuel mileage dropped. Our consumption of gas rose with the size and number of vehicles. Then hurricanes Katrina, Rita, and Wilma hit the Gulf of Mexico and the United States. Besides flooding New Orleans, they wrought havoc with the oil rigs in the Gulf, disrupting the oil flowing to the United States. Refineries were damaged and gas production was curtailed. This triggered another round of gas price increases. World oil production was unable to fill the gap and oil prices soared above $70 a barrel. Much of the United States experienced gas prices over $3 a gallon. Many people thought it was a temporary increase that would soon fall back to a more normal level, but the problem is much bigger than a temporary disruption.

Then in the summer of 2006 British Petroleum was forced to shut down part of the oil supply from Alaska's North Slope due to corrosion in the oil pipe line that further limited oil from the United States. Other sources were able to make up the differences and oil prices started to decline. But was that a reason to think our problems are over? In the spring of 2007 the price of gas soared again as prices reached all time highs. This time the lack of refining capacity was blamed as rising demand drove the price up. By the spring of 2008 oil prices were over $120 a barrel and gas prices were at all time highs with no end in sight. By Memorial Day 2008 oil prices had reached $135 a barrel and gas was knocking on the door of $4 a gallon. Food prices were soaring. The trucking industry was experiencing many failures. By the 4th of July 2008 oil was over $140 a barrel and regular gas was well over $4 a gallon. People were beginning to realize we were really in trouble with oil.

Let's go back and look at what is happening with oil. Even though the use of oil goes back before recorded history it was not until modern times that it became important. Production of oil was very limited through the nineteenth century. In 1859 William A. Smith sank a well in Titusville, Pennsylvania that produced ten barrels of oil a day that doubled the United States production so that it equaled that of Russia. But it was not until 1901 when the gusher Spindletop in Texas ushered in the dramatic growth of United States oil production. With this discovery in the early part of the twentieth century the United States led the world into the third era of energy ? the era of oil. The United States became the largest oil producer in the world. Then in 1948 the United States went from a net oil exporter to an importer of more foreign oil than it produced. By that time the Middle East oil fields were growing in importance and the largest oil field in the world, the legendary Ghawar, was discovered in Saudi Arabia in the same year that the United States became a net importer of oil.

In 1956 geologist M. King Hubbert working for Shell Oil made a very controversial prediction that US oil production would peak about 1972. In fact it peaked in 1970 and proved that Hubbert was correct. His theory was that oil production would follow a bell shaped curve, starting slowly then rapidly increasing until production could no longer be maintained and then dropping off at about the same rate as it had increased. The key to Hubbert's prediction was making an accurate estimate of the total oil that could be recovered. He was obviously pretty close. The discovery of oil in Alaska and the Gulf of Mexico distorted the curve some but was not enough to change the peak.

His approach is important to understanding what is happening in the world today. In 2001 Kenneth S. Deffeyes, Professor Emeritus at Princeton University and a former geologist at Shell Oil who worked with M. King Hubbert, wrote a book titled: **Hubbert's Peak; The Impending World Oil Shortage**. Deffeyes analyzed world oil reserves and used Hubbert's method to predict world oil production. He called this Hubbert's Peak. His calculations showed the peak at about 2005. There can certainly be local variations in the curve that will move the actual peak, but he did not see a plausible way that it could be postponed beyond 2009. We now have added a new terminology to our language, "peak oil".

These conclusions were based on the world having a little over 2 trillion barrels of recoverable oil with the peak occurring when half of it is gone. Unfortunately, the United States Geological Survey (USGS) prediction is that there is over 3 trillion barrels available, and as Deffeyes says, "that requires discovering an additional amount of oil equivalent to the entire Middle East." At the low rate of new discoveries since 1979, this appears to be improbable. The USGS was predicting that the peak of world oil production will be in 2043. This prediction is probably one of the reasons that our government has done nothing significant to solve our future energy problem.

As further evidence of declining oil production we need to look at the North Sea status. North Sea oil production peaked in 1999. Its decline was then predicted to be 7%, but it has been accelerating from 7% to 8.5% to 11%. And the number of major oil fields discovered around the world fell to zero for the first time in 2003, despite an obvious increase in technological expertise. Another disconcerting loss is occurring in Mexico where their production dropped 20% in 2006 and as one expert remarked its production is now in "freefall". This is particularly serious for Mexico as 40% of the governments revenue comes from oil.

There was the discovery of a new oil field in the deep water of the Gulf of Mexico south of New Orleans in the late summer of 2006. This is estimated to be a large field but when considered in the total picture of world oil reserves will not have a major long lasting impact and will take several years to develop. The other potential oil source in the United States is in Alaska. There has been ongoing debates over drilling in the ANWR area. This area has been projected to contain between 5.7 and 16.0 billions barrels of oil. Even if this large potential was tapped it could only increase world production by 1 to 2%.

The Middle East has been the acknowledged source of vast quantities of oil with Saudi Arabia having the largest fields. They have been looked at as the source that can meet any demand for years to come. However, Matthew Simmons, a respected oil banker recently published his book titled: *Twilight in the Desert –The Coming Saudi Oil Shock and the World Economy.* He analyzed Saudi oil fields and concluded that they do not have the oil resources that they claimed over the years since they nationalized the Aramco Oil Company in 1982. Their fields, though immense, have been producing for over 50 years and are mature fields in decline. They will not be able to increase production to meet future demand.

Saudi Arabia has maintained close control over any information about their oil production and reserves. However, in the June, 2008 issue of National Geographic there was an article titled, "**Tapped Out**" written by Paul Roberts, the author of, **The End of Oil**. In this article he reported the discovery made by Salad I. Al Hasseini, the head of exploration and production for the state-owned oil company, Saudi Aramco in 2000. He had been studying data from the 250 or so major oil fields that produced most of the world's oil. He added in all the new fields that the oil companies hoped to bring on line, and when he added the numbers he concluded that the oil experts "were either misreading the global reserves and oil-production data or obfuscating it." Instead of steadily rising output each year, Husseini showed output leveling off, starting as early as 2004. "Just as alarming, this production plateau would last 15 years at best, after which the output of conventional oil would begin a gradual but irreversible decline."

This was not very popular with the leaders of Saudi Aramco and Husseini is now retired as an industry consultant. World oil production as reported by the Department of Energy, Energy Information Administration is shown in the following chart. The numbers are the Annual Average in millions of barrels per day.

YEAR	2004	2005	2006	2007	2008
Millions Barrels/Day	83.10	84.58	84.54	84.59	85.47

Annual Average World Oil Production – From: Energy Information Administration

As you can see worldwide production has been essentially flat for the last five years, with the highest peak in 2008 by a small margin. With the world currently in recession 2009 will probably be down, so 2008 could turn out to be the world peak oil year.

Dubai in the United Arab Emirates is spending vast sums to build an upscale tourist center. They are doing this so they will have a source of income as they know their oil will run out in about ten years.

Even with Saudi production in question, the Middle East has about 67% of known reserves with the rest of the world the remaining 33%. This includes such countries as the United States, Venezuela, Russia, Canada, Mexico, China, and several other smaller producers. In 2005 the non Middle East producers were producing about 57% of the total daily world consumption.

The amount of oil reserves is only half of the issue. The other half is demand. The United States currently consumes 25% of the daily world use of oil. China has now become the second largest consumer and its demands are rapidly growing. China is projected to soon have as many automobiles as the United States. India is currently the number six consumer with a rapidly growing demand. Worldwide population growth is also driving an increase in demand in many other nations. This rapidly growing demand is compounding the problem of the oil resources left in the world.

Many people do not believe that "peak oil" is real. They are crashing ahead with their lives and driving their gas guzzling SUVs thinking someone will discover a vast new source of oil. Some point to the tar sands in Canada that is a huge source of oil and the oil shale in Colorado. Certainly Canada is extracting a significant amount of oil from their tar sands since the price of oil has risen so high, but the penalty to the environment is large and if the oil shale in Colorado is exploited the problems escalate dramatically. Can our world absorb the environmental impact? Later chapters will look at this subject in greater detail.

The issue of when the oil peak is reached is actually not the most important issue. The real problem is that oil is a finite resource and it will be depleted at some point. If the peak has actually occurred or if it occurs within the next few years we will suffer a traumatic economic dislocation as energy prices rise to unbelievable levels as people and nations compete for the oil that is available. As we have seen it only takes a few percent difference between supply and demand to trigger large increases in price. History has made this clear. The world is already feeling the pressure of increased food prices and even shortages. The use of corn to make ethanol has driven the price of corn sky high and taken much of it out of the food stream. Rice became scarce enough that the large outlets in the United States were limiting the number of bags to each customer when they had any to sell.

Alternative energy systems cannot be developed fast enough to prevent an economic dislocation from happening. It takes 30 to 40 years to develop a vast new energy source. Even if the peak is when the USGS predicts, we have to start right now to avoid the disaster! But the evidence that is building in the fall of 2008 indicates we are already on the verge of an economic catastrophe. The melt down of banking and financial institutions that occurred in September of 2008 and the stock market fall in October were only precursors of what would happen.

The other reason that we will experience a huge economic recession or depression is because of the enormous impact the cost of energy has on every aspect of modern life. It is not just the cost of driving our cars. It is the cost of the food we eat which is raised with machinery that uses energy, fertilizer that is made from petroleum, energy to process it, trucks or trains to transport it, and the cost of heating and lighting our markets. The costs of all the products we buy are made with the use of energy and they must be transported to the stores and our homes. Heating our homes takes energy. Some homes burn oil, others use natural gas or propane. These fuels will experience similar cost increases as gasoline. Energy is involved with every aspect of our lives. People will not be able to afford to buy new cars. We will be forced to curtail driving our cars as fuel costs escalate as demand exceeds supply. Airline travel will be too expensive for most people. Many airlines have already added charges for checked baggage and all food and drinks in an effort to cover their fuel costs without dramatically raising rates. Soon they will be forced to raise fares. Consumer spending will drop when people do not have the money to buy nonessential items. Jobs will disappear. Unemployment will soar and great numbers of families will be reduced to poverty levels. The only good thing that can come out of this is that demand for oil will fall and therefore reduce the magnitude of the price increases. As happened after the oil embargo of 1973-74 demand could increase again driving prices back up. We will see these price oscillations, but each swing will result in ever higher prices. In the meantime the cost in human suffering will be devastating.

It has been three quarters of a century since the start of the great depression and not many people are still around that remember how devastating it was to live through that period. Today there are different forces at play in our world, but they have an even greater impact on our lives that will lead to a similar global economic dislocation. When we add the problems we face with energy, the future is grim indeed, unless we move rapidly to find a replacement for oil.

• • • • • • • • • • • •

Energy is so vital to our civilization that nations do not hesitate to go to war over it. I will never forget a December day in 1941 when I was a ten year old boy playing with my friend that I learned that fact. My friend's dad was building a garage and had the radio on as he was digging a trench for the foundation. The program he was listening to was interrupted and the voice on the radio reported that Pearl Harbor in Hawaii was under attack by the Japanese. Ships were burning and sinking. The air field was being bombed and strafed. We were now in World War II even though it had not yet been declared. The Japanese had attacked Pearl Harbor to destroy the United States Pacific fleet to protect their oil shipping routes from Indonesia which the US had threatened. So we were drawn into World War II over oil!

As a kid it was terribly exciting and we played war in the trench of the garage foundation while we listened to the attack described on the radio. It was only later the terrible implications of what was happening sank in.

My older brothers, who were in college or off working, came home a couple of days later to announce they were joining the military. My oldest brother joined the Navy and became an officer on a minesweeper that later participated in all of the invasions of Africa, Sicily, Italy, and Southern France. He was then transferred to the Pacific in command of a minesweeper and spent the last 18 months of the war leading in the invasion fleets as they fought their way towards Japan. Another brother who was already a lieutenant in the Army Reserve couldn't pass his physical for active duty and spent the war working in the military aircraft industry. My third brother joined the Navy as a medical corpsman and was later transferred to the Marines. They all survived the war, but their involvement drove home to me and our family the terrible nature of war.

When we were attacked on September 11, 2001 our world changed. The attack came from al Qaeda under the leadership of Osama bin Laden. He was from a Saudi family and many of his followers were from Saudi Arabia. They were angry about the United States presence in the Middle East and our support of Israel. With a base in Afghanistan they drew the focus of the American effort to capture or kill bin Laden . As a result of the war the Taliban that controlled the country was overcome and a new government installed. Unfortunately, Osama bin Laden was not found and is still free. Our focus on Afghanistan was lost when we turned our attention to Iraq.

We attacked Iraq and its rich oil fields. It is a war we started with a pre-emptive strike under the guise of stopping a dictator who supposedly had weapons of mass destruction that he could use to attack us. The case was presented to the people as our fight against the terrorists and to keep the United States safe. As we now know there were no weapons of mass destruction and there is no evidence of their cooperation with al Qaeda. Even so the war has become, "a war on terrorism." Attacks on Western nations were not originating in Iraq. We now

find ourselves in a quagmire in Iraq with no real good resolution. We have imposed our ideas of democracy on a people that do not want them. As outsiders we simply do not understand the culture of that part of the world. Many more Iraqi civilians have been killed than if we had left Saddam in power.

Undoubtedly, part of the real reason we went to war in Iraq was because of oil. They have large reserves of oil that are not being tapped to their fullest and are under attack by the insurgents. In addition, corruption siphons off significant amounts from the current government. Iraq has become a training ground for the insurgents that use terrorist methods. The war is costing this country a huge amount of dollars and thousands of American lives. In 2009 the end is finally in sight. But it was a war that divided America as did the Vietnam War.

Iraq is not the only war in the Middle East. The conflicts between Israel and the Palestinians has smoldered for years. The conflict between Israel and Hezbollah in Lebanon is part of the hate that permeates the people in that part of the world. Now there is Iran which is aggressively pursuing nuclear power and the refining of nuclear material that could be used in bombs. Besides, the West can not afford to lose their large reserves of oil.

We have lost the respect of the world with our actions in the Middle East, but with most of the world's remaining oil reserves being in that area we are inexorably tied to this unsettled part of the world. We cannot live with them in peace and we cannot maintain our current standard of living without them, as long as we depend on oil for energy. The Middle East is not our only problem. Venezuela's President Hugo Chávez uses his oil wealth to promote socialism and put the squeeze on his ideological enemies. When the competition for oil rises to a high level in the world, will we be able to deal with him? I think not. So we face not only the issue of depleting world reserves, but where they are located. If we look at the true cost of oil, we must add in the cost of the wars that are occurring and will expand in the future. The prospect is grim.

● ● ● ● ● ● ● ● ● ● ● ●

I am not sure just when I became aware of the fact of global warming. But as I was doing the research for my first book the subject kept cropping up. At first it was thought to be a problem suggested by a few crazy scientists that was without merit. How could we possibly be responsible for upsetting the balance of nature on a vast scale? The idea was ridiculed by most of our political leaders and a great number of our people. However, the evidence continued to accumulate that it was in fact happening and we were responsible. The problem that was identified as the major cause was the burning of fossil fuels. Many scientists still didn't accept this theory or didn't think it was real. When the Kyoto Protocol was put together in an attempt to attack the problem by trying to reduce the amount of carbon dioxide and other greenhouse gases, our leaders took the stance that it would hurt our economy and refused to sign the agreement. One of my scientific friends that read my first book in 1995 tried to convince me that carbon dioxide couldn't cause global warming. But years have passed and the evidence has been building.

It is a phenomenon that has happened before. Scientists have now discovered that 250 million years ago there were massive volcanic eruptions that covered a third of the land area of Siberia with lava. This caused enormous quantities of carbon dioxide gas to enter the atmosphere. This great increase of carbon dioxide resulted in the rapid warming of the earth. The result over the next 10,000 years, or less, was the extinction of 80% to 90% of all life on earth. The evidence in the rocks indicates that this has happened at least 11 times in the last 250 million years with a great death of species recorded.

The frightening aspect of our current situation is that at the rate we are adding carbon dioxide to our atmosphere, the world will be at the same level of carbon dioxide in 100 years that caused the catastrophes in the past. If we continue business as usual until 2050 we will be past the point of being able to stop the massive damage that will occur. The evidence continues to build that this is a real phenomenon. All the glaciers in the world are receding. The ice in the polar caps is melting. In September of 2008 a Manhattan-size ice shelf broke loose from Ellesmere Island in Canada's northern Arctic, another dramatic indication of how warmer temperatures are changing the polar frontier. The vast amount of ice that covers Greenland is melting at an increasing rate. If only the ice in Greenland were to melt completely it would raise the water level in the oceans nearly 20 feet. This would flood low lying nations and destroy many of the world's port cities.

In the June, 2007 issue of National Geographic Magazine the featured story was, "The Big Thaw". Its sub heading stated, 'Its no surprise that warming climate is melting the world's glaciers and polar ice, but no one expected it to happen this fast." It described the process of the melting of the ice on Greenland and how the process is being accelerated by the feedback loops in the process. It describes how melting water flows down through the ice and lubricates movement of the ice towards the sea where it is melted by the warmer sea water that causes the rapid calving of ice bergs. The companion stories describe the rapid melting of the polar cap ice and the loss of habitat for the creatures that form the basis of life for whales, seals, and polar bears. The extinction of these mammals will be placed at risk before the end of the century. It is a sobering prospect.

Burning fossil fuels is not the only cause of global warming. Deforestation is another culprit. The Amazon basin in Brazil is in its second year of extreme drought. About a fifth of the Amazonian rainforest has been razed completely and another 22% has been harmed by logging, allowing the sun to penetrate the forest floor, drying it out. When you put these two numbers together they come close to 50 percent. This has been postulated as the "tipping point" that could mark the death of the Amazon. If, or when, this happens it could start the release of the 90 billion tons of carbon dioxide that is stored in its forest. This is enough to accelerate the rate of global warming by 50 percent. There is evidence that deforestation has led to the demise of other societies in the past. It is possible that with the rainforest gone the Amazon could become a desert, accelerating global warming further.

Scientists now agree global warming is a real phenomenon that has frightening consequences for our future.

• • • • • • • • • • • •

The problems we face are compounded by the earth's exploding population. Through the millennia of man's early history the population grew very slowly as disease, wars, natural disasters, and food sources limited population growth. But with the development of civilization and the ability to produce more food and control disease, the population growth has been dramatic. Just 200 years ago the world population was less than a billion people. A hundred years ago it was only

1½ billion people. Today it is over 6 billion and growing at the rate of one quarter of a million people a day.

On October, 17, 2006 the United States reached a population of three hundred million people. It only took 39 years for us to go from two hundred million to three hundred million people.

In addition to the rapidly expanding world population, there is the situation of the emerging nations. China with its 1.3 billion people has the fastest growing economy on earth. They have passed Japan as the second largest oil consumer after the United States. Their streets and highways are already clogged with automobiles instead of bicycles. They cannot build power plants fast enough to satisfy the demand for electric power. China is currently adding a new coal fired power plant every 5 to 7 days! India's economy is also rapidly expanding, and their need for energy is growing. They also have other pressing problems, such as the lack of adequate water supplies and sanitary facilities. The need for fresh water is not confined to India alone. Many other nations are also short of fresh water. Considering that the United States currently consumes 25% of the world's oil usage with only 5% of the population, what will happen when the energy requirements of emerging economies such as China and India use the same amount of fossil fuel energy per capita as the United States? The future looks bleak indeed.

China and India are only two of the developing nations. Many more undeveloped nations exist in poverty, unable to provide even the basic infrastructure for their people. They cannot emerge from their current state without energy to fuel the engine of progress needed for a developed society.

As world population grows and as the economies of developing nations expand this places an ever-growing demand on world resources. Where is the limit when the earth can no longer support the increase?

What I have been covering in this chapter is the reality of the problems facing our world and the United States as we look toward the future. But, gloom and doom is not the primary subject of this book. Rather I intend to lay out a path that we can take to solve most of the problems we face. It will take dynamic leadership of our President, Congress, and industry. It is not an easy path. It will take courage and determination to accomplish, but the rewards will be great. It will require the cooperation of many of the other nations of the world. The problems are so large that we cannot solve them by ourselves. However, as you will see, if the United States does not take the lead in this venture, and lets another nation develop the next energy source, we will no longer be the strongest nation on earth. Our children will inherit a far different country that we have let become a footnote in history.

We must have an Administration

and Congress that has the vision and leadership ability to solve our energy problems. The election in November of 2008 was the first step as Barack Obama was elected President of the United States. He was elected as the candidate of change that brings buoyant new life into the future of America. He faces great problems as the United States and the rest of the world slide deeper into recession. But his first moves as he prepares for the transition of power have brought a sense of hope with his selection of strong cabinet leaders and others that will be leading the nation in the coming years. He may not yet know about the potential of solar energy from space, but he does understand the importance of addressing the problems of impending oil depletion, global warming, and wars over energy. We as a people have chosen wisely. Now we need to support our new president as he leads the nation forward in a quest for the solution to the problems we face.

So what is the solution? We must develop an energy system that can replace oil and the other fossil fuels as the primary source of energy. A solution that will eliminate the pollution of our atmosphere, free us of dependency on foreign oil, and provide the ample low-cost clean energy that the developing and undeveloped nations need to emerge to a better standard of living. This energy source is solar power satellites. Solar energy generated in space on giant satellites that will deliver electricity to the earth 24 hours a day to any nation on the earth. The technology is known, but the will has been lacking. It is now time that we stop wringing our hands and go to work to bring in the fourth energy era, solar energy from space which can solve our current world problems. The United States as the largest consumer of energy on earth has the responsibility to lead the way.

2   The Energy Eras

Energy holds a unique position in the development of civilization, because it is energy that makes it possible. Without energy we wouldn't be able to cook our food, heat our homes, travel great distances, run machinery, have light at night, communicate with the rest of the world or enjoy many of the conveniences that make our lives what they are today. The first humans that lived on our earth had very limited use of energy. They had the sun and the strength of their bodies. The sun has been at work from the beginning of the solar system. It supplies the light that allows plants to grow through photosynthesis. It provides the heat that warms the planet sufficiently to support other life forms. It supplied the energy to grow forests, evaporate water into the atmosphere, and along with the rotation of the earth invigorated the atmosphere to develop wind and weather and storms. By the time humans arrived on the scene the sun had provided the energy needed to grow the life forms that through the ages were transformed into oil, natural gas, and coal.

From this foundation early man started the first energy era. It was the era of wood. There were plenty of trees and occasional lightening storms provided the spark to start fires. Man could now cook food, stay warm on a cold night and even keep the wild animals at bay during the nights. The challenge was to keep the fire going since early humans had no way of starting a fire themselves. If the coals went out they would have to wait until the next lightning storm. At some point along the way they discovered that they could start a fire by rubbing dry sticks together. It was much, much later after the discovery of iron that the use of iron and flint made life easier. The spark of iron on flint would readily ignite dry tinder and later on gun powder. In 1827 John Walker discovered sulfur friction could also start a fire and he developed what became known as "Lucifer" matches. Flint and steel became relics of the past.

Wood was an ideal source of energy for early people since it was available nearly everywhere and gathering it was easy. Fire made it possible to make pottery and to live in the higher latitudes. The era of wood lasted for uncounted millennia and provided the basis for the development of our civilization. It was the energy of burning wood that provided for the advancement into first the Bronze Age starting about 4000 BC and then the start of the Iron Age about 3000 BC. Without the ability to work metal into tools and implements, our civilization as we know it could not exist.

Wood wasn't the only source of energy, but it was the major source. People had the strength of their muscles and supplemented these with domesticated animals to multiple their effort. Wind was put to use to power ships sailing the seas and to power windmills that drove milling machines. Wind was the first energy source besides animals that was able to multiply the ability of people to do work and travel long distances on the seas, but it was only a supplemental source of energy which came mainly from wood. Light was provided by candles and oil lamps.

Throughout the era of wood civilization expanded and spread with the help of the wind, implements of metal and the warmth of burning wood. The Northern latitudes were settled and populations grew as industries evolved. England had become a significant part of the developing world and was using wood to fuel their manufacture of iron utensils, tools, and weapons. Wood provided the warmth in their homes. Wood was used to build their ships. It was used in the false work needed to build their cathedrals. The demand had grown to such an extent that wood was imported from mainland Europe and from their colonies. Great quantities of Kauri trees from New Zealand were cut and shipped to England. These trees supplied the material for the masts and rigging of their ships. The harvesting of the giant Kauri was so complete that they stripped New Zealand of nearly all of their Kauri forests. Kauri grows to be giant trees that rival the Sequoia in size and live for over 2000 years. Today there are only a few remaining in New Zealand where they are now protected.

This was typical of what was happening as demand increased and could not be satisfied from their own lands. The cost rose as wood had to be imported from outside of England. The first energy crisis was underway.

It was a crisis that had no clear beginning and no clear end as there was still plenty of wood in the rest of the world. But by 1580 the capacity of wood could no longer meet the needs of the expanding nation of England. They even banned new buildings in London to restrict growth of the city. It was the beginning of the end of the era of wood, which had started when humans harnessed the first fire. The era of wood is still not over for some of today's poor nations that have stripped their land of trees in an attempt to stay warm and cook their food because they do not have the resources to pay for coal or oil.

In other ancient societies wood was not only used for cooking and warmth, but also for other purposes. On Easter Island the trees were essential for making rope from the bark and for moving and erecting the giant stone statues that ring the island. The trees were also cut to provide for farm land to raise their crops. This had the disastrous result of allowing erosion to wash away the fertile soil. In the end all the trees were gone and there were none remaining to build canoes for fishing and for commerce with other islands. It left the people stranded without the ability to fish or to raise enough food to survive. Their civilization collapsed. This was not the only civilization that deforestation destroyed, but it is a startling example of what did happen and can happen again. Today the only trees on Easter Island are the few that have been planted in recent years. Now this is happening in the Amazon and if this continues it will impact the entire world.

• • • • • • • • • • • •

England was fortunate in that they had coal, an alternative source of energy to tap. The second energy era, the era of coal started in England as a result of their shortage of wood that forced them to look for and use an alternate source. At first it was used to supplement wood and then to nearly replace it. It could be picked up off the ground in some places. Then the first mine was established at Newcastle in 1233. Coal started to replace wood for home heating, but iron was still made using charcoal made from wood. It was not until 1640 that coke was first distilled from coal and from then on the iron and steel industry blossomed as new processes were developed to use the enhanced properties of coal.

When the steam engine was invented in 1769 it gave humans the ability to greatly multiple the amount of work they could accomplish. Now one man running a machine could do the work of ten men or a hundred, maybe even a thousand. The magic that enabled this explosive development was energy from coal. Other nations jumped on the band wagon as well, but it was England that led the way up through the nineteenth century.

Coal fueled the Industrial Revolution. Its partner was the steam engine burning coal, which brought the Industrial Revolution to its full development. England, a small island nation, emerged as the dominate economic force in the world. Its empire grew to cover the globe. Coal was king and England made the most of their opportunities.

• • • • • • • • • • • •

It was not the shortage of coal that started the third energy era, the era of oil. It was the enhanced capability of oil that propelled it to the forefront of the energy world. Oil and its refined products of gasoline and diesel that could be easily transported and used in minute quantities to fuel internal combustion engines, made it an extremely useful fuel. Oil had been known since early history, but it was not until the start of the twentieth century that the era of oil began with the discovery of oil in Texas in 1901. Oil became plentiful and cheap. It could do many tasks better and cleaner than coal. As a result it soon replaced coal as the fuel of choice.

The ability of oil to be refined into gasoline and diesel allowed the rapid development of the automobile as the means of personal transportation. This development alone changed the entire complex of civilization. Then came airplanes and the world was opened. Sailing ships were first replaced by coal fired steam engines and then by oil powered engines that made worldwide commerce practical.

During the first half of the twentieth century the United States was the largest supplier of oil on earth and it elevated the United States to the forefront of economic development in the world, displacing England. The lessons are clear, whatever nation develops and controls the world's leading energy source dominates world economy.

We are still in the era of oil as no other energy source has emerged to take its place, but its time is limited. We are fast approaching the time when it can no longer satisfy demand. In addition the United States is no longer in a position to control the oil that remains.

As in the era of coal and now the era of oil they are not the only energy sources in use, however, they are the primary sources. Wood is still being burned in parts of the world as an energy source and in developed areas for its convenience and to give us pleasure. Coal is used extensively today to generate electricity. Hydropower is used wherever possible. Nuclear energy is used to varying degrees around the world. Natural gas is used to heat homes, for cooking, heating water and to generate electricity. Alternative energy systems have been developed to supplement our other energy sources in some countries, but the magnitude of their contribution has been limited in most cases.

Today our civilized high technology world depends on energy for its very existence. Energy makes it possible to grow enough food to feed a world of over 6 billion people. It provides the productivity to supply the material and machines for our modern world. It has given us computers and technologies that were unimaginable just a few years ago. It powers our automobiles, trains, ships, and airplanes that give us mobility to travel for work or pleasure. We now live in a global community. Energy has given us the ability to go to the far reaches of space.

What will happen to our world when we do not have enough energy? That is the question with a very grim answer. A more important question is: What is the energy source for the future? A source that can supply the world with energy that can replace oil and move us into the fourth energy era.

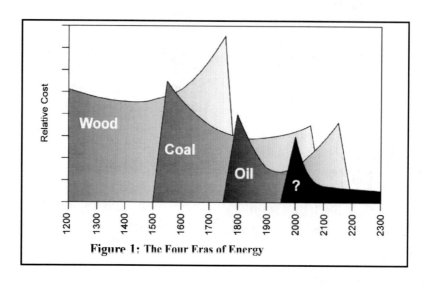

Figure 1: The Four Eras of Energy

If we choose correctly the fourth energy era can last as long as there are people on earth for the sun has been the primary source of earth's energy from the beginning of time. What we need to do is tap into it directly.

In order to answer the question of what that energy source should be, we must establish the criteria for an energy system that can fill the requirements for our future. When I was Boeing's Program Manager for Solar Power Satellites from 1975 to 1980, one of the first things we did was to establish a set of criteria that our system needed to fulfill. Criteria are like the rules of a football game. If you don't have rules to play by the result is chaos. It is the same for developing an energy source. As I mentioned in my first book, **SUN POWER**, the Department of Energy laughed when I showed it to them and said they didn't need any criteria, other than it couldn't be expensive. "We will know what energy system to work on when we see it". You know the results of that position. Nothing significant has been accomplished.

The criteria we established then are still valid today. Five very simple criteria, if fulfilled, will address the problems we face today. Even though the criteria are simple it does not mean they are easy to meet. In fact they are very difficult to fulfill. They are:

1. **Low-cost**
2. **Nondepletable**
3. **Environmentally clean**
4. **Available to everyone**
5. **In a usable form**

These five requirements cover all of the critical aspects of a future energy system but do not define a specific solution. The test of their validity is in their inherent logic and in the authenticity and comprehensiveness of the data used to develop them. The criteria are derived from what history has taught us about bringing prosperity, preserving the earth, and minimizing world tensions. Each broad criterion addresses a generalized goal. As potential solutions are identified, the criteria can be subdivided into more specific requirements that will help us evaluate how a solution might satisfy the broad, general criteria.

With the criteria defined, it is possible to measure how a known energy source might satisfy the requirements or might point in the direction of a new solution. The criteria may appear simple, but they make the task of finding a suitable energy source for the future extremely difficult. When a way is found to satisfy all of them, the benefits to our country and the world will be well worth the effort. An energy system that can satisfy these criteria will form the basis for the fourth energy era. While not entirely replacing all other energy sources, it will provide the core and help keep the price of the supplemental sources within reason.

Let us look at each of the criteria in more depth and see how they can help lead us to a solution. The first is: **Low-cost**. The way the Department of Energy looked at low-cost was to consider the system that could provide the first kilowatt hour of electricity at the lowest cost. This automatically eliminated any large scale system that required development. They were not prepared to make major investments in our future. Low-cost as applied to this set of criteria is the ultimate cost of the energy that is produced. To give you an example, in the early part of the nineteenth century oil in Pennsylvania was selling for two dollars a gallon. That would be equivalent to $47 a gallon today. When we consider that we are looking for a vast source of energy it only makes sense to believe it would take a large investment to develop it. So this criterion is primarily directed towards the cost of energy after it is developed and being used in great quantities. Over the long term it is the cost of the energy that counts not how much it costs to develop. One of the reasons that nuclear power has not expanded in the United States, besides the worry of nuclear accidents, is because of the low-cost of natural gas generated electricity. Natural gas was so cheap that nuclear power couldn't compete. This illustrates how cost is a major driver.

The second criterion is: **Nondepletable**. This criterion rules out all stored fuels because stored fuels are finite by nature and all would eventually be depleted if they were used in large quantities. Both coal and oil are examples of finite stored fuels. It took uncounted millennia for these fuels to be formed by the growth of the life forms that provided the hydrocarbons that natural forces ultimately converted

into coal and oil. This is a process that takes an incredibly long time. In the case of oil we will be using all of the recoverable oil in about two hundred years. Coal reserves are larger, but its contribution to atmospheric pollution is particularly bad. As a result, we need to consider a system that is either continuously replaced or one that is vast enough to last indefinitely. This exemplifies the process by which the criteria can guide us toward a solution without inhibiting the reasonable possibilities.

The third criterion is: **Environmentally clean**. This is an extremely important criterion. We now know that our world is experiencing global warming. If we want to have any hope of reversing this process we must dramatically reduce the burning of fossil fuels that feed carbon dioxide into the atmosphere. This is not the only environmental problem we face, but if we want our world as we know it to survive we have to stop the pollution. Our energy source for the fourth era must be a clean, nonpolluting system.

The fourth criterion is: **Available to everyone**. There are several reasons why this is an important criterion. Wars have been fought over energy. As I mentioned Japan attacked Pearl Harbor in 1941 in order to destroy the American fleet with the purpose of protecting the sea lanes between Indonesia and Japan so that they could maintain the flow of oil. Iraq invaded Kuwait in 1990 to gain control of their oil fields. Ample low-cost energy could provide the path to the development of the poorer nations that have very limited access to energy now. The developing nations such as China and India are building their development on energy. Energy is absolutely essential to their development as it was to the United States.

The last criterion is: **In a useful form**. The first three energy eras were based on hydrocarbon fuels. During the era of wood the energy form was basically heat with a byproduct of limited light. In order for wood to supply sufficient heat to process iron it first had to be converted to charcoal. When it was used to make steam for a steam engine its range was limited by its heat limitations. The era of coal introduced much higher heating capacity that is still being utilized today in coal fired electric generating plants. Oil provided a major leap in usefulness with its ease of transportation, ability to fuel engines directly, be an efficient heat source and do a variety of other tasks including generating electricity. It is electricity that has the ability to do nearly any energy task. It is electricity that powers our industries, lights our homes, enables our communication systems, made computers and the internet possible. It can be used to separate hydrogen from water for mobile fuel. It can provide the energy to desalinate sea water to fresh water. It can be used to charge batteries to power our tools and ultimately our cars. Electricity is the most versatile energy form that we have. It is a candidate for meeting the fifth criterion.

As we approach the end of the era of oil there are many alternative systems that have been investigated to some degree. Some of these systems are in use today and others are being developed. But none of them have emerged as a clear winner to become the energy source for the fourth era. They all have some shortfalls in meeting the criteria that is so essential for the system to replace oil. The energy source that I am proposing has not been developed because of the magnitude of its development costs, political pressures, and lack of vision. However, the one energy source that can meet all of the criteria for the fourth era is solar energy generated in space on solar power satellites that would transmit the energy to the earth with wireless power transmission beams. I will show you how it meets all the criteria and how it will solve many of the problems we face today, if we have the will and vision to bring it into being.

3   Solar Power Satellites Begin

When the 1973-74 OPEC oil embargo jolted us into the realization of our dependence on foreign oil for our economic livelihood, our country started a frantic effort to investigate alternative energy sources. The first government agency to address this problem was the Energy Research and Development Administration (ERDA). In 1977 when Jimmy Carter became President he proposed the formation of the new Department of Energy (DOE) which absorbed ERDA and was formed around the Atomic Energy Commission (AEC), which also included responsibility for developing atomic bombs. It must have seemed logical to the administration at the time to group atomic bombs with energy. We were still in the midst of the cold war. I suppose they didn't trust the Department of Defense to have control over developing such powerful weapons. In any event, the effect of the Atomic Energy Commission inclusion has distorted the effort of DOE ever since. As one senior official in DOE told me one day, "If you look at what the Department of Energy really does, the last thing you would conclude is it develops energy systems." However, they are the government agency responsible for our government's energy programs.

The first major thrust of the government was to emphasize energy conservation. This included reduced speed limits, lowered thermostat settings in government offices, mandatory fuel mileage for new automobiles, stringent insulation requirements for new construction and higher efficiency goals for hot water heaters and refrigerators. There were also campaigns to encourage the populous to conserve energy wherever possible. This was the right response to solve the immediate problem. Conservation is the only thing that can be done in the short term.

President Carter was an avid follower of the "small is beautiful" approach to developing distributed alternative energy sources that could be developed in small scale by individual home owners. This emphasized solar applications, such as solar hot water heaters, passive solar home heating, individual home solar panels, and many other ideas that sprang up along the way. Some of these ideas were installed on the White House roof during this period. I had occasion to brief one of Carter's lawyer friends from Atlanta who sat in many of his cabinet meetings as an observer. I briefed him on the potential of solar power satellites and one of his questions was; "Will it run the air conditioner in my new house? Jimmy, doesn't want me to install a bunch of stuff that would take a lot of electricity, but I need an air conditioner."

Carter's background in nuclear submarines also gave the nuclear advocates in the Department of Energy a free hand in promoting nuclear power research to the detriment of other large systems, which Carter opposed. This period did see the start of terrestrial solar generating systems development that included concentrating mirror systems, fields of solar cell panels, and wind generators. In addition many other ideas were pursued, that included ocean thermal gradients, geothermal, ocean waves, and tidal generators. Some of these ideas have evolved into useful energy sources.

While we were sitting in line to gas up our cars, cursing OPEC and wondering why our government didn't do something, the concept of solar power satellites moved from a science fiction dream to a solid candidate for becoming the energy source for the fourth energy era. In 1968 Dr. Peter Glaser of Arthur D. Little Company first proposed the concept of a solar power satellite. NASA started low level studies of the concept in the early 1970's before the oil embargo to determine if they were feasible. This was when I first learned about them and realized that they would become the focus of my professional life.

It was two years after the embargo before I was able to join the program in 1975 and became Boeing's Solar Power Satellite Program Manager. During this time NASA expanded the studies in partnership with ERDA. In 1977 DOE took over with NASA retained as the technical leader. In 1977 we bid on and won one of the two Solar Power Satellite Systems Defini-tion Contracts sponsored by the Department of Energy and NASA. The contract we received was very comprehensive and covered all of the systems required to design and develop the satellites, including the space transportation systems. In addition, Boeing provided me with matching funds for research and development of complimentary technology.

A parallel contract was won by Rockwell International. They had the same work statement we had so we were in competition to develop the definition of the system. However, following NASA's policy we attended each others briefings. Both prime contractors were supported by subcontractor teams that included the very best of the aerospace technology companies in the country. Fred Koomanoff was the DOE Program Manager. He had put together a comprehensive program that in addition to the System Definition contracts, included a large number of other contracts that investigated other aspects of the systems, such as the environmental impact, societal impact, and comparative costs.

The system we were contracted to design was for giant satellites placed in geosynchronous orbit that would provide 5 gigawatts (5,000 megawatts) each of electrical energy to the earth. The memories of my youth surged through my thoughts because what we were doing was developing a Grand Coulee Dam in the sky. Each satellite would have nearly the same output as Grand Coulee Dam.

The satellites would be in geosynchronous orbit, 22,300 miles above the equator, where they would be in sunshine 24 hours a day for over 99% of the year. They only passed through the shadow of the earth during brief periods before and after the spring and fall equinox when the earth's axis is perpendicular to the sun. As a result they could deliver energy 24 hours a day for nearly the entire year. This is the great advantage of going to space and meant that the sunlight shining on the satellite was five times more than the best location on the earth and 15 times more than the average location.

The satellite we designed was covered with solar cells that converted the sunlight into electricity which was then routed to a phased array transmitting antenna, similar to a giant radar antenna. The transmitter converted the electrical energy into a radio frequency wireless power transmission beam that beamed the energy to a receiver on the earth that rectified the radio frequency energy back to DC electricity. This was processed by inverters into AC electricity that was fed into our normal electric distribution grid. All of the technology needed for the satellites was based on known principles. The solar cells we proposed were single crystal silicon cells similar to the ones being used on communication satellites of that period and terrestrial solar panels. The difference was the cells we were proposing were thinner in order to save weight. They were only 2 mills thick and had an efficiency of 16.5%. We had a solar cell company manufacture cells for us to verify their characteristics.

The wireless transmitter design was based on the work done by Bill Brown of Raytheon. He demonstrated the concept in 1964 when he flew a model helicopter powered with a wireless power transmission beam. He had succeeded in transmitting energy without wires and had finally fulfilled Nikola Tesla's dream of a half century before. Early laboratory tests demonstrated 54% efficiency of the beam from electricity in to electricity out. Later tests by NASA at the Goldstone facility transmitted 30 kilowatts of energy over a mile to a rectifying receiver at even higher efficiency.

The biggest challenge to the system design was the immense size of the satellites, twenty square miles of solar cells and a transmitter that was one kilometer in diameter. This meant that all of the hardware had to be transported to space and be assembled in orbit. We needed a railroad to space like the one that came through my home town hauling hardware to Grand Coulee Dam. Unfortunately, we couldn't build a railroad, but we could design a space transportation system that could do the job.

In the late 1970's the only launch vehicles in use were rockets that were launched into space and thrown away after each flight. This resulted in very high launch costs. It was obvious to us that we must have a reusable system that could be used over and over like a commercial airliner or cargo plane. So the design of a new fully reusable launch system became an integral part of the system definition. In addition we developed assembly techniques that were amenable to robotic assembly with a minimum of manned assembly.

The progress made during these studies was spectacular. We were able to define the system in great detail. The satellite system though enormous was simple in concept and its efficiency was high. It became clear to us that solar power satellites could become the energy source for the world that could replace oil and stop atmospheric pollution. We could wean ourselves away from Middle East oil and provide energy to the developing nations to allow them to evolve to a higher standard of living. It looked like the cost of electrical energy could be reduced to levels similar to hydroelectric dams.

During the period the studies were underway there was a lot of public awareness of the program and the attitude of most people was, "why don't we get on with it?" I spent a lot of my time making presentations around the country to publicize the potential. This drew the ire of the nuclear and fossil fuel advocates in the administration. The nuclear fusion proponents in the Department of Energy saw the rapid progress we were making on the satellite system and knew they had to stop us. They realized that if solar power satellites were developed it was the end of nuclear fusion. They simply could not compete on an equal basis so they set out to kill the program. The administration was now playing hardball. The president of Boeing Aerospace Company was warned by the government to shut me up or Boeing would suffer the consequences.

This was the foretaste of what was about to happen. As the studies reached their conclusions political pressures built. When we assembled for the final reports we were informed that the program was to be cancelled and we were warned that **no reports were to be released to the pubic**. The government slammed the door on the most important program of our time. If we had gone ahead with development in 1980, we would have been well on our way to having an energy source to replace fossil fuels. We wouldn't have had to go to war in the Middle East because of oil. We would have a clear path to stopping global warming and the space frontier would be open to all.

The public never knew why the program disappeared. Now 29 years later there are very few people who have even heard of, or remember anything about solar power satellites and what they can do for the future of the world. Even today the oil interests dominate government policy and our thinking about energy. They will use every tactic imaginable to try and stop development of energy from space, because it is the one energy source that can eclipse oil. It will take massive public support to overcome the roadblocks raised by big oil.

The nuclear fusion advocates are still around, but they are no longer an effective voice as there has been so little progress towards development of practical nuclear fusion power plants.

• • • • • • • • • • • •

The ending of the program was the start of a dark period in my professional life. Ever since I had made the offhanded comment as a teenager that I would one day go to the moon, my life had been focused on space. Even in retirement, energy from space is still the goal of my life. Its achievement had been clouded and I was without a compass to steer by. I applied for and was granted a 6 months leave of absence from Boeing to reassess my future and what I was going to do. I spent the time working on our boat and then a long cruise into Canada to clear my mind and plan for the future.

I returned to Boeing shortly before Christmas of 1980 as Proposal Manager for NASA's Solar Electric Propulsion Stage (SEPS). This job was short lived as about two months later we were notified that funding for the program had been stopped. I was again at loose ends. There wasn't any way I could restart solar power satellites. I was assigned to a small group working on new space programs. Our boss, a retired Air Force Colonel didn't seem to know what we ought to be doing, so I went looking for something back in Washington DC. What I found was the High Frontier organization. It was a nonprofit group led by retired Army General Daniel O. Graham. He was proposing that the United States should develop a space based ballistic missile

defense system and also was interested in solar power satellites. By this time Reagan was President and Graham had good contacts within the White House and with the Secretary of Defense.

Graham had a small group of people at High Frontier, but the real work was being carried out by volunteers from many different places and organizations. Graham asked me if I would help them out. When I asked my boss if it would be OK, he said, "yes", because he didn't have anything else for me to do and it would get me out of his hair. So I picked a couple of my guys and we set out to help Danny Graham put together a space based ballistic missile defense system. It focused on using a constellation of low earth orbit satellites that carried a complement of small rocket interceptors. These interceptors did not carry any warheads. They relied entirely on the kinetic energy of impact to destroy a ballistic missile or its atomic warhead while in space.

Unfortunately, we also had scientists from the nuclear weapons research facilities involved and they wanted to include the potential of adding lasers and atomic weapons to the mix. The most troubling idea came from Ed Teller who proposed we should use X-ray lasers that required an atomic weapon to initiate the laser. We had several heated arguments over the subject. In the end the proposal that was sent to the Secretary of Defense and the White House was based primarily on kinetic weapons with a mention of lasers.

To do the cost analysis of the system I brought in one of the economists who had worked with us on solar power satellites. It came out looking quite good. I volunteered to put the final report together and print a number of copies for submittal to the President. So I bundled all of our material together and headed back to Seattle to assemble the report. When it was completed and printed I made another trip back to Washington DC and delivered the completed report to General Graham to take to the White House. During all of this time I had given my boss at Boeing briefings as to what we were doing, but I don't think he really understood what was happening. So after I gave General Graham the copies of the reports and returned to Seattle I decided it was time to make my boss fully aware of what was going to happen if President Reagan bought into our proposal. The reason it was so important to Boeing was they were the developer and producer of most of the United States ballistic missiles that were about to be made obsolete and a new industry was going to be formed around the space defense initiative and I knew what it was going to look like!

My boss finally understood what he had let me do. He turned white as a sheet and stammered as he told me to put what I had told him in a letter and hand carry it immediately to the Boeing Space Division Vice President's office. I was being sent to the principle's office. I did what I was told and as I handed the letter to the secretary I told her she should take it right in to her boss. I returned to my office on the first floor to find a message that I was to call Boeing's Corporate Headquarters Office immediately. The way Boeing was organized started with the corporate officers at the top, with several sub-companies with their own presidents reporting to Headquarters. I was in the Space Division of the Boeing Aerospace Company. When I returned the call it was from the office of the Boeing Company President, directing me and the economist that worked for me to be at the corporate board room at 2:00 PM and not to talk to anybody in the meantime.

I had hardly hung up the telephone when I heard my name screamed from the upstairs office of my division's Vice President. My letter had reached the upper echelon. I did not respond to the division Vice President's summons, but instead rounded up the economist to go to headquarters. That is when I found out why they wanted to see us. The economist had spent some time at corporate headquarters before he came to work for me and was a friend of one of the corporate Vice Presidents. He had run into him that morning and told him what we had been doing. The V.P. understood instantly what that could mean to Boeing. So when we walked into the Board Room it was packed with all the senior leaders available in Seattle from all of Boeing's major companies, including Commercial Airplanes, Aerospace, Military Airplanes, and Technology. The President of Boeing at the time, T. Wilson, was out of town so his next in line was in charge.

He directed us to brief them on what we had been doing and what we thought was going to happen. I gave them the whole story and told them I expected that President Reagan would act on our input because both he and the Secretary of Defense were quite enthusiastic about the concept. Our briefing and discussion went on for over an hour and then we were directed to wait outside while they considered what to do. We weren't sure if our heads were on the block or if we were going to have new jobs.

Then they brought us back in. A task force was being formed under the leadership of a Senior Vice President that specialized in space technology. He was to have access to the very best people from anywhere in the company. The two of us were to be part of the task force. It would meet in secret at an isolated location and we were not to tell anyone, including our bosses where we were or what we were doing. Our task was to analysis the potential capability and the cost of a space based ballistic missile defense system and report back to headquarters with our findings.

What followed were several weeks of intense activity. At the conclusion we briefed headquarters on our findings that concluded the concept was feasible and the costs could be held within reason. Boeing headquarters said, "We can not keep these results to ourselves but we must brief the Department of Defense". This was done. It was not very much later that President

Reagan announced his Space Defense Initiative, which unfortunately got dubbed Star Wars and was later distorted by the Air Force towards lasers. But it was one of the key elements that brought down the Soviet Union and ended the Cold War and I was proud to be a part of it.

The Boeing Aerospace Company President at this time had built his career on ballistic missiles and other missile programs and he held me responsible for the business that was about to go away. He ignored the opportunity to prepare for a business that was literally dropped in his lap. His animosity had another unfortunate aspect for me. He forced me out of the Aerospace Company of Boeing. The Space Defense Initiative was the result of the time and effort of two of his company's managers and Boeing did not take advantage of the information. It was only later that Boeing realized their mistake tried to get back in the game.

• • • • • • • • • • • •

My recourse was to call an old friend who was President of the Boeing Military Airplane Company and ask him if he had a place I could hide. So I disappeared for the next couple of years on a black program that today is know as the B-2 bomber. It was not space and I hated it, but it was a job while I looked for another opening in the space program. My break came when another old friend called to say NASA wanted to re-compete the

External Tank for the Space Shuttle because of dramatically rising costs of the tank contractor Martin. The Space Division brought me back to evaluate the potential, so I was back in the space program.

After that effort was completed I spent the last few years of my working life developing the design concept for a fast response reusable space transportation system for the Air Force, later I moved on to Boeing's effort to develop the National Aero-Space Plane. It was exciting to design an airplane that could fly all the way to orbit. I was in charge of the structural design, an extremely challenging effort. By the time I retired from the program and Boeing at age 55 it was apparent that the technology for the engine was not far enough along for it to be successful.

So for the next 6 years I set aside my dreams of space and enjoyed the life of cruising the island nations of the South Pacific with my wife and our cat in our sailboat. It was a wonderful time that came to an end because of the call of space. It happened while we were visiting a friend in Sydney, Australia having left our boat moored in Brisbane. I read an article in the Sydney Herald about a conference that had occurred in Paris on solar power satellites. It quoted my old friend and colleague Lucien Deschamps who was the leader of the French solar power satellite activities. All the old excitement and dreams came tumbling back. I was on an airplane the next morning to return to our boat with the goal of writing a book advocating the development of solar power satellites.

It was a tumultuous period of writing, trying to find an agent and wondering how to get the boat back to the States. We had to leave the boat in Australia for a year while I went back to the States to work on the book. As it took shape I realized we had to give up our dream of sailing around the world and return home.

As all of this was going on I completed writing **SUN POWER: The Global Solution for the Coming Energy Crisis**. It required a major adjustment to be back working on promoting the satellite energy system after six years relaxing in the paradise of the South Pacific. To promote solar power satellites I formed a small company, Solar Space Industries and we were able to interest Boeing in updating the data base on solar power satellites. Part of the contract from Boeing was the task of preparing briefings for the Department of Energy and NASA. The results were interesting. The Department of Energy maintained their stance of disinterest, but NASA responded by initiating the "New Look Studies" of solar power satellites at a low funding level. Unfortunately, the Boeing Vice President that had contracted our studies retired and his replacement, a former Air Force General, didn't want anything to do with solar power satellites so Boeing's support for the effort was terminated.

As the years passed I despaired of

ever seeing the development of solar power satellites in my lifetime. I thought I had accomplished all I could and was resigned to letting others carry the banner. Then the unbelievable happened. I received an e-mail message dated January 3, 1998. It was from a European investment group interested in developing solar power satellites and asked for my collaboration with their plan. They said they had read my book, SUN POWER. I thought someone was pulling my leg, but decided to play along. I answered the e-mail and told them I was interested and exactly what did they want to do and who were they? The e-mails moved back and forth as little by little they explained who they were and what they wanted to do.

For over two years we corresponded and met together as the plan developed. They proposed we form a joint venture to develop solar power satellites as a commercial venture without any government funding. They would be responsible for arranging the funding for the system and my company would be responsible for the design, development, and assembly of the satellites. After researching the potential world energy sources and the people involved with them they reached the conclusion that the solar power satellite system was the only system that could meet future demand and that I had the best approach for their development. As a result they came to me and Solar Space Industries to undertake the program. This was a startling proposition as the company had been

inactive since 1995. We didn't have any employees and the people that were working with me promoting the concept were all volunteers. Nevertheless the opportunity was too great to ignore. If they could put the funding together we could put the team together to develop the system.

In our first meeting with the group from Italy we asked them why they wanted to develop solar power satellites. One of the senior members of the group gave their answer in a short impassioned speech. The essence of his statement was: "We have to do this for the future of the world, the people of the third world nations have no hope without energy and this is the only source big enough to meet their needs and our needs. Because oil is controlled by a few nations it is going to lead to war and we want to maintain peace. It is not for the money we are doing this."

In order to establish the funding required for the program it was necessary to update the design concept and prepare a preliminary plan for implementing the program. The starting point was the design developed by the DOE/NASA Solar Power Satellite System Definition Studies conducted in the late 1970's. This had been updated by the work of Solar Space Industries in 1993-95. To this was added the technology advances made over the years since the original studies along with information gathered from NASA's New Look studies. Since the program was not to be government funded with tax dollars, but rather a commercial venture, all the costs needed to be accounted for and balanced against the revenue the system would produce. The costs needed to include development, the entire space infrastructure required, and building the satellites. High initial costs had stopped the program in the past. Now we had to satisfy the harsh reality of a commercial venture.

They wanted to purchase 44 five gigawatt output satellites. So we swallowed hard and put together the estimates for both development cost and the cost of building the satellites. This was an immense challenge with a funding requirement to match. I started by proposing a conservative schedule that held initial funding to a minimum to prove all the key elements worked together before committing to full-scale development. They did not like that approach at all. They wanted to develop and deploy the satellites in the shortest time possible. This was essential to them as I found out later, because of the high interest rates they were going to be charged by the international banks who were going to supply the funding. There would be no looking back, failure was not an option. We reworked the schedule to the shortest time we could and gave them the plan. It was accepted.

We signed a Memo of Understanding as an interim agreement while they worked to obtain the necessary commitment of funds. They expected to have the funds committed in six months. We didn't see how they could possibly raise the monies that were required, but that wasn't our part of the job. We immediately went to work to prepare for the biggest commercial program ever attempted.

Six months passed and in the end they were unable to secure the funding commitment, it was just too huge. But as the time passed I had learned more about their overall plan for raising the money. It was extremely innovative and came within a hairs breath of working. They had put together a consortium of international banks that were going to back the sale of bonds to raise the necessary funds. But the real clincher was going to be the hostile take over of a major international corporation that would provide Solar Space Industries with a built-in industrial base and additional development resources. Disagreement over this issue was what caused the international banking consortium to break up. However, it changed many things for me. It renewed my effort to initiate a space solar power development program. It established the interest of the financial institutions in investing in solar power satellites that opens new avenues for program development. And it clearly brought home the fact that other nations of the world are dead serious about developing space solar power and recognize its critical importance. It should be a wake-up call for America.

• • • • • • • • • • • • •

It was not long after the European effort collapsed that I received another interesting telephone call. Gene Meyers, told me about an idea he had about using Space Shuttle elements to launch commercial space stations and potentially solar power satellites. When I learned more about his company, Space Island Group, and what they were planning, I was a little gun shy after my recent experience, so was a little skeptical. Nevertheless, I agreed to co-operate with their efforts.

The appeal of a totally commercial program is very strong, but I felt it must be tempered with reality. The immense size of the program and its broad international implications makes it imperative that there is some government involvement. Previously, I had developed a program that combines government and industry into a partnership that would be a composite of the best of both worlds. In the late summer of 2000 I was called to testify before the House Congressional Subcommittee on Space and Aeronautics, concerning solar power satellites. I presented an outline of this proposal of a government/industry partnership in my testimony on September 7, 2000. It is a plan that will form the framework for what I propose in this book to develop the system.

It is time for the people and the leadership of the United States to step forward and face the challenges of the new century. At the top of the list is energy. Without ample low-cost energy many of our other problems are unsolvable. The energy crisis before us will have a devastating impact on our economy and standard of living if we do not act. Short-term solutions by themselves will only delay the inevitable, even though they are necessary to bridge the gap until a new source can be established. This is a call-to-arms for the American people to focus our effort on the long-term solution to our energy future, space solar power.

Fortunately there are still a few people around who are working to bring the program back into public awareness. It is essential to bring new, young talent into the fray as those of us who worked the program twenty-five to thirty years ago are gradually slipping from this world. Bill Brown, one of the giants, is no longer with us. Most others, like myself, are retired but still trying. We need help.

• • • • • • • • • • • •

There are some emerging organizations that are attacking the problem. Since that call several years ago when Gene Meyers told me about his vision, Space Island Group has developed a unique two part plan that solves the problem of the initial high cost of space transportation to launch solar power satellites.

Their plan follows the path that Gene originally described to me of using the elements of the Space Shuttle system to create a new launch vehicle. The design would eliminate the Space Shuttle orbiter and add the Shuttle orbiter engines to the external tank. The solid rocket boosters would be retained and in place of the orbiter another external tank would be added which would be converted into a space station. A crew transport vehicle would be added on top of the space station modified external tank. Both external tanks would be launched into low earth orbit. The one that was converted into a space station on earth would be ready for use by commercial companies to lease for research or production. The one that had carried the hydrogen and oxygen for the launch would be empty and ready for conversion in space to another space station. This vehicle would also have the capability of launching about 200,000 pounds of payload.

The idea is a little like Skylab, America's first space station. It was built inside of an empty third stage from the Saturn V rocket. The new part of the system would be a crew module mounted in front of the empty hydrogen tank. It would be patterned after the DC-X concept developed by McDonnell-Douglas a number of years ago. It used a conic reentry shape and would land under rocket power on its base. In the Space Island Group configuration it would be able to carry 30 personnel to orbit and back.

Space Island Group's concept was attractive to potential users, but has not yet provided sufficient interest for potential financing sources to advance the development funds. Now with the second part of the plan to launch solar power satellites with the launch costs being paid for by leasing out the commercial space in the space stations, the plan becomes attractive to the potential users of electricity from space. This solves the launch cost hurdle for the initial solar power satellites and provides for the revenue stream that will come from selling electric power on the earth. Several countries, including Japan, China, and India have expressed interest in purchasing the power. If a purchase agreement can be reached the program would be underway. With the start of the program the interest level would raise and provide the foundation for the necessary investment for developing the reusable space transportation system that is required for the large number of launches to provide the of satellites that are needed to supply the world with clean low-cost electricity.

In addition to Space Island Group there are other organizations that are emerging as awareness of the potential of space solar power becomes more wide spread. With the election of Barack Obama as President with his commitment to develop renewable energy sources, the future for space solar power could be very bright indeed.

4   Our Energy Situation Today

Natural Gas is one of the key sources of energy in use today to heat our homes, cook our food, generate electricity, and with many other uses. It didn't start out that way. In many of the oil fields natural gas was an unwanted by-product which was burned off in giant flares that were clearly visible from early space flights. At least the product of burning gas was less harmful to the atmosphere than just releasing the gas into the atmosphere, since the natural gas is a much worse greenhouse gas than the resulting carbon dioxide that was produced. As time went on natural gas has become a useful fuel and is now contributing to the world's energy needs. As it was with oil, the United States is a major producer of natural gas. Kenneth S. Deffeyes book, **BEYOND OIL: The View from Hubbert's Peak**, has an excellent chapter on natural gas. As Deffeyes points out it is more difficult to estimate the amount of gas available than is oil, but as he also says in his book concerning United States production, "between 1980 and 2002, the best of the natural gas targets were drilled. We're now being served leftovers."

Matthew R. Simmons, in his book, **TWILIGHT in the DESERT: The Coming Saudi Oil Shock and the World Economy**, discusses the production of natural gas in Saudi Arabia for their own use. In Saudi Arabia natural gas is used to generate electricity, desalinate sea water and to support the petrochemical industry. The demand for natural gas has grown rapidly with their exploding population. However, the search for natural gas is getting more and more difficult and expensive. New gas wells are dropping off peak production very soon. So even in Saudi Arabia, natural gas production will probably not be able to support their own needs much longer.

In the United States we have been using natural gas to satisfy our growing demand for generating electricity. It now generates 16% of our electricity. Unfortunately, we are no longer able to produce as much as we need and are importing natural gas from Canada. Some natural gas is imported by ships that carry the gas as liquid natural gas (LNG). Because of the cold temperatures required the tanks are insulated and under pressure. Also, because of the danger of transporting LNG there are only four ports in the United States that can handle LNG carriers. They require that all other ship traffic be suspended while the ships enter the port.

There has been much speculation about the abundance of methane hydrates as a source of energy. There are apparently vast amounts of it in the ocean bottoms, in oil and gas fields and other places in the ground, but the problems of extracting the methane gas are large. So far there has not been much success from the effort.

One of the desirable features of natural gas is that the level of carbon dioxide released in the atmosphere when it burns is less than oil or coal. Nevertheless the use of natural gas does contribute to the problem of global warming and if we are to solve the problem of global warming it can not be considered as a candidate for the fourth era, even if we unlock the secrets of extracting methane from the methane hydrates. All the other signs of natural gas availability indicate it does not have the capacity to fill the growing needs of the world.

• • • • • • • • • • • •

Coal was once the world's dominate energy source. It slipped from that place because of the superiority of oil to fill our needs, not because we were running out of coal. When I was a kid growing up, our home had a coal furnace and it was my job to shovel all the coal to feed the furnace. It didn't take long after I went off to college that my dad converted the furnace to oil. There are large reserves of coal in most of the industrialized nations and it is used extensively to generate electricity. There is enough coal in the world to last for quite a long time, but the punishment to our environment is devastating. China is the largest producer and consumer of coal in the world. Currently China is adding one coal fired power plant every 5 to 7 days to their generating capacity. Their use of coal has become so large that they have now become a net importer of coal with the majority of it coming from Australia. It

is projected that their coal production peak will be reached about 2020. Their pollution filled atmosphere is the result. Coal has been used for other forms of energy at different times in the past. Germany processed coal into liquid fuel in the later years of World War II, because they were cut off from their source of oil. South Africa did the same during the embargo in the Apartheid era.

Currently 50% of United States electricity is generated with coal. This is down from 56% a few years ago because of the rapid growth of natural gas generating plants. The total amount of electricity generated by coal is still increasing. However; it just isn't growing as fast as natural gas.

As Deffeyes said in his book, **Beyond Oil**, "Coal is the best of fuels; it is the worst of fuels. Best, because it is much less expensive per unit of energy. Worst, for a long list of reasons: killer smog, acid rain, atmospheric carbon dioxide, mercury pollution, acid mine drainage, and a choice between hazardous underground mines and surface-disturbing open-pit mines."

There is much talk about "clean coal," but I think the term is an oxymoron. There isn't much clean about coal, even when it refers to sequestering the carbon in the earth instead of allowing it to become carbon dioxide released into the atmosphere. The cost of doing the sequestering is apparently high and I do not know how efficient the process can be, but the chances of it becoming a universally working scheme is extremely doubtful.

As we move into the Fourth Era, coal should be the first energy source to be displaced because of its environmental problems.

• • • • • • • • • • • •

There is no energy source that stirs up emotions and controversy like nuclear power. 19% of our electricity comes from nuclear plants, but it has been many years since any new plants have been built in the United States. Nuclear accidents, nuclear waste, and cost overruns have chilled the atmosphere in this country to the point of suspending new construction. This may change as the cost of energy ramps up and the realization of global warming permeates the population.

Nuclear power is the primary source of energy in France, now providing 79% of their electricity. But, there is a strong movement by the Green Party in Germany to close their nuclear plants. Japan surprisingly has a significant number of nuclear plants and adding more, but they have had several minor nuclear accidents that have created a anxious atmosphere. Iran is seen as a threat to the world as they work to develop nuclear power and maybe nuclear weapons. North Korea is another nation that is escalating the threat of nuclear weapons as they detonated an underground nuclear device and continue in their development of nuclear technology. Some recent negotiations could possibly lead towards suspension of their nuclear weapons efforts.

All of the commercial nuclear power plants are fission reactors that use uranium as their fuel source. The designs vary, with some being safer than others, but they all produce radioactive nuclear waste that must be stored or handled in some way to keep the radiation isolated. The Bush administration and other advocates of nuclear energy were calling for its expansion to help fill our energy needs. It does have the advantage of not producing carbon dioxide gas as a by-product of electricity generation. Its environmental problems are focused on nuclear waste and how to store and dispose of it. In addition there is the fear of accidents and nuclear proliferation by using the technology and reactors to make bomb grade materials. The availability of uranium fuel is also an issue, if the world was to use it as a primary energy source. The availability of the fuel can be greatly expanded if the plants were to be developed as breeder reactors. The down side of this approach is the resulting fuel is plutonium, great for making bombs. We already have an unstable world. Nuclear proliferation could lead to the end of civilization. Already there are many nations with nuclear weapons capability that include Pakistan, India, Israel, Russia, France, England, the United States, and now North Korea and Iran working to join them.

The other potential nuclear energy

source is fusion power. This has been the dream of many scientists for the last 60 years. Fusion is the combining of atoms whereas fission is the splitting of atoms. Our sun is a fusion reactor. Hydrogen bombs use a fission bomb to trigger a fusion reaction. In theory fusion reaction does not have the radiation danger of a fission reaction. The problem is developing sufficient pressure and temperature, to first trigger, and then maintain, a controlled reaction. Through the years uncounted billions of dollars have been spent trying to develop fusion power. It is still an unfulfilled dream with no clear path to success. If the money spent on fusion power in the 1980's had been applied to solar power satellites, the computer I am using to write this book would be powered with solar energy from space.

I am sure that as we move deeper into the energy crisis we face, nuclear energy use will be expanded. It will be used by the countries that feel comfortable about it. The cost of the power plants will play a significant role in how many are built. Several years ago in the state of Washington, Washington Public Power Supply System (WPPSS) planned to build five nuclear power plants. Only one plant was ever completed and put into operation. The others were abandoned as costs escalated. If energy costs from nuclear power plants can be made competitive they will be built as long as the nuclear fuel is available. The cost of energy from nuclear power needs to include the cost of finding a solution for handling the nuclear waste they produce and the cost of decommissioning the plants. So far that has not been done and there is not a permanent solution in use. The Department of Energy has prepared a site in Nevada for the permanent storage of nuclear waste, but so far political resistance has kept it from being activated.

• • • • • • • • • • • •

One of the first sources of commercial electricity was demonstrated when Edison designed the first hydroelectric power plant that was built at Appleton, Wisconsin in 1882. This was followed by many others around the world. The largest is the Three Gorges Dam in China. It will have a generating capacity of 22 gigawatts. Hydroelectric energy is one of the cleanest power sources we have. It is renewable energy that does not pollute the atmosphere and has no cost for the fuel. However, it does require disturbing the natural flow of rivers and inhibits fish migration. A major social issue is the displacement of huge populations as happened in China. There are some good benefits such as the formation of lakes and reservoirs, but on the other hand some scientists now worry about possible local weather effects.

The construction of Grand Coulee Dam during the depression not only created desperately needed jobs, but also raised the economy of the Northwest by providing huge amounts of low-cost electricity and an irrigation system that turned a desert into rich farm land. Electricity generated during

World War II was vital to the production of aluminum for building the airplanes needed to win the war. After the war the cheap clean power from Grand Coulee and the other hydroelectric power plants on the rivers of the Northwest has sustained a dynamic economy in the region.

In balance it is probably the best energy source we have today. Unfortunately its capacity is limited and there are very few rivers left to dam. Existing dams will be able to supply energy far into the future, unless global warming changes our climate to the point where the rivers stop flowing. But even without this potential, we will not be able to increase its capacity significantly to meet growing need.

• • • • • • • • • • • •

Conservation is not an energy source in the strictest sense, but it can be used to dramatically reduce the amount of energy we use. The danger is if it is the only approach and is carried too far. Drastic conservation can inhibit the development of civilization and prevent the emergence of the developing nations which need energy to evolve. Conservation has been described as the organization of scarcity. However, if done correctly it is like free money. Conservation becomes mandatory when the cost of energy increases beyond our ability to pay for it. We can reduce our travel, turn to car pools or mass transit, and get rid of our gas guzzling vehicles. We can reduce our thermostat settings, use less hot water, pay more attention to the lights we use, drain our hot tubs, and learn to better organize scarcity. High fuel taxes forced travel conservation in Europe and much of the rest of the world. Their tax structure has kept the cost of fuel in most countries higher than what we were paying in the United States with oil at $140 a barrel. In these countries a majority of vehicles are small high mileage cars. For example the streets of Italy are packed with Smart cars that get 65 miles to the gallon along with many motor bikes and scooters. In much of Europe 35% of the cars are powered with high mileage diesel engines.

The American automobile manufactures are already experiencing huge loses as people are buying Toyotas, Hondas, Hyundai, and other high mileage cars. General Motors and Chrysler are begging for government bailout money and are on the verge of bankruptcy. American manufacturers did not learn their lesson very well when the same thing happened after the 1973-74 oil embargo. They may not be able to survive this time as the crisis is not going to go away. It will be interesting to watch what happens at Ford, where Bill Ford Jr. stepped aside as CEO to bring in Alan Mulally to turn the company around. Mulally was brought in from the aerospace industry and has no automotive business background. Being an outsider may be what it takes to understand what is happening with energy and turn things around.

Conservation can reduce the

amount of energy we use in many ways. The key to saving energy without reducing our standard of living is through increased efficiency and use of the natural environment. On the island where I live we have many people who are passionately concerned about the environment and what is happening in our world. As a result we have formed a renewable energy co-op with the goal of reducing our use of energy and to develop renewable energy sources on the island.

One of the first efforts of the co-op will be to perform energy audits on homes of residents who want to participate. The goal is to identify how the homes could be made more energy efficient. As part of our early research about what can be done, we were given a tour of a home on the island that had been built for a retired engineer as what he called "a solar home." It was a lovely home, certainly not a compromise in livability. They also had a home in the Seattle area that was a typical house about the same size. They spend about half the year in each home. The one on the island uses water catchment from rain water on the roof as their source of water that is stored in large tanks on the property. The catchment water system uses a filter and treatment system that ensures its purity. The heating system starts with a sun room that provides warm air that is circulated through a cement-lined crawl space under the floor that heats the house. When this is not warm enough it is supplemented with a heat pump that has its pipes buried 4 feet underground

where the earth temperature averages 55 degrees Fahrenheit year around. The hot water tank is heated by another heat pump. In addition skylights are used extensively to reduce the need for lights during the day. The plan is to add solar panels to help supply electricity, but they were not yet installed. Even so the net result is the cost of energy to run the home on the island is one fourth of the cost of the energy for the home in Seattle. This is an example of what can be done to reduce our use of energy.

As we move past peak oil production, conservation will be the primary way of mitigating the rising cost of energy. It will become a way of life.

• • • • • • • • • • • •

Biodiesel and ethanol both have origins in plant life and face similar limitations in their ultimate use. On the plus side the carbon dioxide released when they are burned is balanced by the carbon dioxide they absorbed when they were growing as plants, however, their processing takes energy. Also on the negative side they use land that could be used to grow food for a hungry world. In the case of Brazil which produces a large quantity of ethanol the problem is compounded because they have cut down vast areas of rain forests to use for growing plants as well as for the lumber. The net result of that happening may precipitate the demise of the Amazon rain forest which would be devastating for global warming and world climate.

We have seen how deforestation has been one of the primary causes of the collapse of societies in the past. It is a danger that must be avoided at all costs if we are to survive on this planet.

The use of biodiesel and ethanol can be a positive situation as long as their production is limited to plant growth that is surplus to the need for food or use plant products that otherwise would be discarded as trash. Recycling vegetable cooking oil to biodiesel is an effective use of a waste product, but it is a limited resource.

In the meantime, many new start-up companies are springing up around the country to produce bio-diesel or ethanol. These companies can be started with a relatively small financial investment and farmers welcome the opening of new markets for their products. However, availability of resources will ultimately limit these fuels to a relatively small contribution to our energy needs. We need to be careful that our enthusiasm for these fuels does not carry us beyond the natural limitations of available land.

There is one source of biofuel that may have great potential. That is algae, better known as pond scum. It could be grown in ways that would not compete for land that could be better used for food and it can apparently be tailored to provide a high level of output. One potential location for its production would be under the rectennas of solar power satellites.

In the case of ethanol made from corn, the price of corn has doubled and the domino effect will bring higher prices for beef and pork since corn is a major feed stock for these animals. In addition it raises the cost of food made from corn. The amount of corn going into ethanol in 2008 is now as much as is consumed as food by humans. In addition the amount of energy required to make ethanol is about equal to the energy it provides. Ethanol from food crops is not the answer to our energy needs.

• • • • • • • • • • • •

Windmills have been around for a long time. I can remember the wind mill behind our house when I was a kid that pumped water from our well into a tank at the top of the pump house. This was before our town had its own water system. Windmills have come a long ways since then. They now come in huge sizes and generate significant amounts of electricity in some parts of the world. The United States started the movement, but less than 1% of our electricity is generated with wind turbines and a growing concern over bird kills is slowing their deployment. The main concern is focused on their use in the paths of migratory birds. This can be mitigated to some extent by selecting sites which are away from migratory paths and using installations that raise the height of the units to be above the paths. There are also objections to the visible blocking of desirable views. However, in Europe their use is greatly expanding. Tiny Denmark now generates 20% of their electricity with wind and is the world's major supplier of wind turbines. Wind is one form of solar generated energy that does not stop when the sun goes down. The rest of Europe is also rapidly expanding their use of the wind. The United States is far behind.

This may change in the near future if T. Boone Pickens is successful. He has started a massive campaign in the United States to change our energy status. He recognizes that we must replace foreign oil as our primary source, even though he is one of the great oil men of America. He believes that wind power can make a great contribution as an alternative energy source and is leading the effort to dramatically expand its use in the United States.

The size of the turbines is becoming immense. There will soon be turbines that generate 5 megawatts of power off the coast of Germany. Their blades will be two hundred feet long. As a diagram in the August 2005 issue of National Geographic illustrated, if one was mounted next to the statue of liberty the bottom arc of the propeller would clear the statue of liberty and the top of the arc would be nearly 600 feet in the air.

The cost of electricity from the wind is probably the most competitive source of renewable terrestrial solar energy at the present time. It has the advantage of having no fuel cost, so the cost of power is a function of capital cost, maintenance, and operation. The use of wind turbines will certainly grow throughout the world as the cost of oil and gas rises. The limitation will be finding enough appropriate sites and balancing that with the environmental drawback of bird kills and visual pollution. Wind power can be a major contributor to our energy future, but not large enough to fill the needs of the fourth era.

• • • • • • • • • • • •

As with the other alternative renewable energy sources terrestrial solar power use is growing, but not at the same rate as wind. It has two problems. Cost and the fact the sun goes down at night. In addition many parts of the world have cloud cover that further diminished the output. On our island there are a number of people who live without being connected to the electric power grid. They use solar as their primary energy source, sometimes supplemented with wind and small diesel generators. They must have batteries to store the energy for use at night and when the wind does not blow. They usually have some solar system for heating water and rely on wood for winter heat. They use the diesel generators as the source of last resort. Their lives are simple compared to all the conveniences most of us enjoy.

Our renewable energy co-op encourages the expanded installation of terrestrial solar panels for individual homes and are planning some large community installations. However, these are primarily planned as supplemental energy sources with the homes acting as part of the electrical grid. During periods of sunshine the power will feed into the grid. If the electricity generated exceeds the use, the meters run backward and the power is available to anyone on the grid. When it is less than the usage, our homes use the power from the electric utility. This method uses the utility as a huge battery and is called net metering. With a sufficiently large array it is possible to bring the cost of a home's electric bill to zero when averaged over a year's time. The capital cost of the solar panels and other required equipment is fairly high and will take a lot of years to amortize. However, as the cost of solar cells comes down the prospects improve.

One new method of generating electricity from terrestrial solar is being developed which uses a concentrating mirror system to provide heat to a sterling engine. It has a tracking system to follow the sun throughout the day. The system uses an innovative design for the sterling engine which results in very high efficiency and a long life.

Developments of this type will undoubtedly continue to make terrestrial solar systems more cost effective.

The people of Vashon Island across Puget Sound from Seattle have established a goal of energy independence for their island. They will be using solar, wind, and other renewable approaches to achieve their goal.

The state of New York is also moving to advance the use of renewable energy systems such as solar and wind with legislation that will vastly improve the conditions for New York for net metering.

When we look at the cost of terrestrial solar as a primary energy source for the world the picture is not promising. If we tried to run an entire utility with terrestrial solar we would need to add storage capability to feed the grid at night, in the cold winter months, and during cloudy periods. This factor alone is a huge deterrent to terrestrial solar ever becoming our primary energy source. If we look at it as a supplemental source that can help supply daytime peak power and as a source that can replace fuel burned during the day it can become very attractive as the cost of solar cell panels are reduced.

However, its ultimate capacity is limited because of the inherent characteristic of the day/night cycle of the earth.

There is an exciting new development in solar cells that will reduce the cost of the cells and make them more competitive. These are thin film cells made from copper, indium, gallium, and selenide (CIGS). Nanosolar, a new start-up company, has developed a method of printing these cells on continuous rolls of metal foil using a process similar to printing a newspaper. They are projecting costs reductions of one fifth to one tenth of current cell costs. A large new facility is being built to produce these cells which are being developed for terrestrial use, but would also be the logical choice for use on solar power satellites.

Boeing is taking another approach to reducing terrestrial solar cell costs. They have developed very high efficiency cells that use concentrators to multiple the amount of sunlight on each cell. A cell that is only a square centimeter can generate 17 watts of electricity. This system is being sold for large terrestrial systems that cuts the cost per watt to about half of other systems. Other approaches are also being developed that will surely reduce the cost and make terrestrial solar much more cost competitive in the future.

• • • • • • • • • • • •

Ocean tides contain a lot of energy. One of the ideas that has been suggested is to select a location with really high tides like the Bay of Fundy and build a dam across the mouth so you could convert it to a hydroelectric power plant. The problems and cost of such an idea makes something like that impractical, but there are some new developments that may turn out to be very practical. In fact one of the utilities in our area has applied for a permit to install tidal current generators. The concept is to place large propellers on the ocean floor in a channel that would drive generators. These would be secured to moorings on the bottom and would turn with the tidal current driving the generators. The propeller blades would be variable pitch with the ability to reverse pitch so they always rotated in the same direction regardless of the current direction. The claim is they turn slow enough that fish could swim through the blade arc without being killed. Apparently a system like this is going to be installed in the river next to Manhattan Island. I understand the generators can be made in different sizes so that they could be installed in as little as five feet of water. There is a lot of force due to moving water. I know by trying to keep the propeller on my sailboat from turning when under sail is no easy task.

If this system really works it could supply a significant amount of nonpolluting power to a grid. There would be down times during high and low slack tides when there is no current. It will be interesting to watch how this develops. The inventiveness of mankind is amazing. Here again is a potential source of energy that can help, but it is not universal as it would need to be located in areas where there are strong tidal currents.

• • • • • • • • • • • •

Another way to generate electricity from the ocean is tapping ocean waves. Studies show that there is more potential for generating electric power than is available from tidal currents. This is because generators could be placed along the entire sea coast where there is significant wave action. The approach is to have moored floats that move up and down with the waves on their moorings. They would work like the hand held battery chargers that charge a battery when you shake them back and forth. There are penalties to consider with these types of generators as they must be moored where there is severe wave action so the environment is pretty hostile. In addition, in order to generate a large amount of power it will require a large number of them that could interfere with boat traffic and sea life. I do not know if there is a good estimate of the cost of power that could be generated in this fashion.

• • • • • • • • • • • •

Geothermal energy, another promising concept with New Zealand, Iceland, and the United States having some geothermal generating capacity. The sites that are generating electricity now are those that have occurred naturally where the geo-thermal activity is near the surface and accessible for development. So far the efforts to develop geothermal energy outside the natural sites have not been commercially practical. And unfortunately, the developed sites are losing output because they are cooled by the process of extracting energy. Without some new development we are not likely to see an expansion of this source of energy.

Other potential sources such as small scale hydroelectric are achieving some success and do provide small amounts of energy along the way, but their total contribution will probably remain insignificant.

Except for bio-diesel and ethanol, what is significant about the other energy sources I have talked about so far in this chapter is that they generate electricity. This is significant because electricity is going to play a bigger role in our lives in the future as we move into the fourth era. I will show how we can use electricity to power most of our world later in the book. In the meantime we need to discuss our current situation with oil.

I have discussed most of the various sources of energy we are using or are developing, but there are others and undoubtedly more in the future, as we search for the right solution for our energy needs.

The one thing that is common to all the energy systems I talked about except for the bio-fuels, was that their form of energy is electricity. What this means is that the energy must be transported from its source to the consumer. This requires an electrical distribution grid. It is becoming apparent that our existing grids are inadequate to handle the demands of the future. There is insufficient capacity to transmit terrestrial solar power generated in the southwest to industry in the eastern part of the country. Wind farms are currently limited not only by lack of wind, but the ability to transmit their output to consumers.

As the new President looks for infrastructure improvements to help create jobs to support the economic recovery, the rebuilding of the nation's electrical distribution grids should be high on the priority list. It is essential to support all of the potential renewable energy systems that are being developed and will also be essential for solar energy from space.

• • • • • • • • • • • •

In the first chapter I talked about the problems we face with peak oil. We can debate when the actual peak will occur, whether in fact it was 2005, or maybe 2008, or even if it is delayed further. But the one issue we cannot avoid is it is a finite resource that is being consumed at a prodigious rate and we will pass the peak of production at some point. The evidence is very strong that it is imminent. We will not know for sure until at least a year after the actual peak. The data available in 2008, shows that 2005 was the peak by a very small margin, as production has been nearly flat since then. In any event oil prices went through the roof starting in 2007 and into 2008 as the supply could not readily satisfy the demand. When this happened the public was forced to limit their use and demand plummeted. As a result oil prices fell dramatically. This pattern shows clearly how closely supply and demand can affect the price.

As we all know, oil is the largest supplier of energy in the world and today's transportation systems are utterly dependent on it worldwide. For-

tunately oil isn't just suddenly going to go away, we will have some time to make changes from oil based fuel to electricity for most applications, but there are some uses such as powering airplanes that have no other options yet, except for bio-fuels. Even so there will be massive shifts in our worldwide industrial structure. Many mature old businesses will simply disappear. History records that most established companies cannot make a major shift in direction and are replaced with upstart companies working in the new emerging field.

The automobile industry is particularly at risk. General Motors and Chrysler are on the verge of bankruptcy and only a government bailout may keep them alive. In any event not many names we know are likely to survive. Most will simply continue what has worked for them for decades. But the day will come, and soon, when it won't work anymore and they will be gone. Just like the Baldwin Steam Locomotive Works that led the world in the manufacture of coal powered locomotives until the day nobody wanted steam locomotives anymore. There will be a massive shift in the paradigm of the world as we move from oil to another source of energy. The old way will be replaced with the new. There will be many losers, but there will also be many winners that grasp the new opportunities. Which group will you belong to?

Oil companies and oil users will fight to the end to maintain their status and continue to search for more sources of oil. They will extract the heavy oil from the tar sands of Canada and Venezuela. They will go after the oil shale of Colorado and drill more exploratory wells in the deep water offshore. This will go on as long as governments allow the continued rape of our environment. But we cannot let it go on much longer if we are to prevent the destruction of the planet through global warming. Humanity is killing our earthly home. The first moves to stop this from happening was when the new Interior Secretary in the Obama administration, Ken Salazar, scraped leases for oil shale development on federal land in Colorado, Utah, and Wyoming in February of 2009. This follows his halting the leasing of 77 oil and gas drilling parcels near national parks in Utah.

So our situation with oil is complex. It is our primary source of energy and supports our civilized world. Its price will go back up dramatically when it can no longer meet demand. It is a major contributor to global warming. Its environmental impact will grow as its extraction from the heavy oil of tar sands and shale are exploited. It is the only known fuel for some of our commerce and it is a primary commodity for the petrochemical industry that supports much of our industrialized civilization.

If we are to maintain our standard of living and provide the energy to allow the developing world to emerge we need to find a replacement for oil as quickly as possible and use the oil that remains judiciously. We have already waited far too long.

● ● ● ● ● ● ● ● ● ● ● ●

On November 19, 2008 James Michael Snead released a white paper with over 100 pages titled, **The End of Easy Energy and What to Do About It.** Before he released the paper and posted it on his web site, he asked me to review and comment on it. I was pleased to do this and found the paper very comprehensive. In it he analyses all of our current energy sources, coal, oil, natural gas and compares their capability to provide for the world's projected energy needs over the next century. He also looks at the capability of nuclear and the renewable sources, such as wind, terrestrial solar, bio-fuels and others. In his analysis he considers the growth of demand that will occur as the earth's population increases and the use of energy per capita goes up as the people in the developing nations strive for a higher standard of living. His assumption is they will not reach the level of the United States, but more like the level of energy use per person in Europe. This results in a staggering increase in the amount of energy that will be required to sustain the world over the next century.

The basic conclusion of the paper is that "Space solar power will be needed to supply most of the U.S.'s and the world's dispatchable electrical power generation capacity while hydrogen produced off-peak space solar power electricity and algae biodiesel will be needed to fill the fuel shortfall." Space solar power is the only source that has the potential capacity to do this.

5   Global Warming

One day in late September, 2006 we were leaving our island by ferry on our way to Anacortes, Washington. It was a beautiful sunny day, warm for September, but then our whole summer had been warm and sunny with no rain. Mount Baker was there in all its glory, but something was wrong. Great portions of it were bare rock where normally there was snow and ice. I was shocked, never having seen it like that before. It just seemed to be one more piece of evidence of our changing climate.

It may not seem like much, stark rocks instead of snow and ice, but we were in for another surprise. As we drove across the beautiful North Cascade highway we were appalled when we entered a vast area of dead forest covered with brown trees. All across our land they are being killed by the Spruce Beetle and Pine Borers. These parasites are normally controlled by cold winters and frosty nights. But now with warmer winters they are totally out of control. At the same time rampant fires are fueled by these dead forests. This destruction can be found in millions of acres of forest throughout the United States and Canada. Can those forests ever come back? How will climate change affect our beautiful country? Some areas may experience what seems like weather improvements to warmer, more comfortable conditions, while others may experience devastating changes, such as colder weather and violent storms. What is becoming absolutely clear is that a worldwide change is happening.

Changing sea temperatures are threatening the growth of coral and the ability of fish species to find sufficient food to survive. Also seafood harvested in the Bering Sea is falling sharply.

On October 1st 2006 the headline in the Seattle Times read, "Ecological upheaval on the edge of the ice." What followed was a story of what is happening in the Bering Sea. It reported, "The Bering Sea is a rich source of America's seafood, most of it harvested by Seattle's fishing fleet. But the fishery is seeing dramatic changes due in part to the shrinking ice cover and higher water temperatures brought on by global warming." It went on to describe the rapidly diminishing numbers of fish and crabs. It described how, "sea birds that once flocked to the region by the millions are in precipitous decline." "The changes coincide with rising water temperatures and shrinking sea ice cover."

"In the Bering Sea…rapid climate change is apparent, and its impacts significant," scientists concluded in the 2004 Arctic Climate Assessment. One more quote from the article sums up the conclusions, "The Bering Sea ice cover pulled back more rapidly than usual this year. For the creatures that are so well adapted to ice — seals, walruses, fish, even birds — scientists have discovered such change can mean a catastrophic upset."

The waters surrounding our island are home to 3 pods of Orca whales. Over the years that we spent sailing in the San Juan Islands and into Canada one of the most pleasurable experiences was watching the Orcas cavort in the waters around us. One memorable day we had friends visiting with us from New Orleans. We were showing them the islands when we happened on a pod of Orcas. We stopped the boat to watch them. As we sat there the number of whales grew and we realized that all 3 pods were there together. They really put on a show, diving under our boat leaping out of the water and tail dancing. Our friends were amazed with a show they had never seen before. It went on for an hour before they moved on. Today if you visit the islands you can take one of the popular whale watching boats that have become a major tourist attraction. But that might not be possible much longer as the Orcas are now an endangered species. Global warming and water pollution are threatening their very existence. The resident Orcas of the Puget Sound region live on Chinook salmon. The loss of wild Chinook salmon runs in the Columbia and Snake rivers due to over fishing and decades of dam building, logging and other salmon-habitat destruction has resulted in reducing the runs to a fraction of their original size. Now global warming is raising the water temperatures of some of the rivers and streams

to temperatures that are lethal to fish. For example the Fraser River in Canada increased about 1.8 degrees Fahrenheit from 1953 to 1998, which yielded a 50% mortality rate among the river's salmon. In the fall of 2006 three Orcas were gone from the Puget Sound pods. It is believed they starved to death.

As Al Gore reported in his book, **An Inconvenient Truth,** the amount of carbon dioxide in the atmosphere in the pre-industrial period was 280 parts per million. When measurements were made from high above Mauna Loa, in Hawaii, starting in 1958 the level was up to about 315 parts per million. By 2005 the level had reached 381 parts per million. The changing slope of the data indicates that the rate of increase is accelerating. This matches our increase in fossil fuel consumption.

In the Arctic the polar ice cap is thinning and the area covered by ice is shrinking dramatically. It won't be long until the ice cap will melt completely in the summer. When this happens the effect on the earth will be traumatic. Polar bears are losing their habitat and have been declared an endangered species. The ice covering Greenland is melting at a shocking rate. Since this is ice that has accumulated on land over thousands of years, ocean levels will rise as it melts. As I mentioned earlier that could be 20 feet, just from the melting ice in Greenland.

To give you a feeling for the effect land based ice can have on the level of the seas, we only need to look back at the last ice age. At that time much of the northern area of the North American continent was covered with glaciers. This ice contained vast amounts of frozen water that had come from the seas. At that time the sea level was low enough that people walked from Siberia to what is now Alaska and continued on to populate the North American continent.

In the Antarctic huge masses of ice are breaking off of the ice shelves. Like Greenland much of the ice there built up over the continental land mass and is about 10,000 feet thick. This ice is the source of atmospheric data that has been traced back 650,000 years by core drilling through the layers that have accumulated over all that time. Air is trapped in tiny bubbles that can be analyzed for its composition.

An article in the February 26, 2009 Seattle Post Intelligencer was headlined, "Ice melting even faster in the Antarctic", was a summary report of a 60 nation team of researcher that has been studying Antarctica for the past two years. They reported that seas could rise 3 to 5 feet. "Glaciers in Antarctica are melting faster and across a much wider area than previously thought, a development that threatens to raise sea levels worldwide and force millions of people to flee low lying areas," the scientists said. The article indicated the big surprise was how much the glaciers were melting in western Antarctica. The biggest of the western glaciers, the Pine Island Glacier is moving 40% faster than it was in

1970s, discharging water and ice into the sea at a much faster rate. The Smith Glacier is moving 83% faster than in 1992. This is happening because the floating ice shelf that normally stops them is melting. They "fear that this is the first signs of an incipient collapse of the west Antarctic ice shelf."

All the glaciers throughout the world are receding at a frightening rate. This ranges from Kilimanjaro in Africa, to the Alps, to Glacier National Park, to the Himalayas, to Alaska and every other frozen part of the world. In addition the permafrost in the Arctic regions is melting. The question is, what does all of this mean to us? Gore pointed out in his book, the Himalayan Glaciers are the source of seven rivers that provides more than half of the drinking water to Asian countries with 40% of the world's population. They are at serious risk with global warming.

Let us consider what will happen as the ice caps of Greenland, the Arctic, and the Antarctic melt and the sea level rises. A great number of the world's major cities and huge population centers will be flooded. Most of the low lying South Pacific Island nations will cease to exist. Much of Florida will become a sea bed as will much of Louisiana and other southern states along the Gulf of Mexico. No sea walls can be built high enough to keep the sea out. Much of the low lying lands of Europe will be part of the sea. Large heavily populated areas in China will be under water. The only good news is that the rising water will come slowly enough that these areas can be evacuated before it happens.

However, there is the possibility that this could happen catastrophically. A number of years ago I read a series of three science fiction novels about Mars, written by Kim Stanley Robinson, that described a disastrous event on earth. It postulated that a vast section of the Antarctic ice shelf slid off into the sea. This caused tidal waves throughout the world resulting in massive damage and loss of life. It was a science fiction story, but with what is happening in the Antarctic today it may not be science fiction after all. There are two different shelves, the West Antarctic Ice Shelf and the East Antarctic Ice Shelf. If conditions deteriorate to the point that they start to melt at their base, one or both could slide off into the sea. The damage and resulting deaths would be unimaginable.

We don't know exactly what killed 80% to 90% of all life on earth 250 million years ago when giant eruptions in Siberia released massive amounts of carbon dioxide into the atmosphere and caused global warming, we only know it happened. Is mankind now capable of avoiding the death rate as the world changes? Are we smart enough to change our ways and stop the cause of global warming? Do we have the will? That is the issue we face.

It is very hard to accept the fact that this time around it is us and not just nature that is responsible for pouring carbon dioxide into the atmosphere. If we can accept that fact we can do something about it. Unfortunately, there are many people and leaders that are still in denial of the reality of global warming. They simply refuse to accept the fact that we can affect the earth's climate and are unwilling to face reality. One email I received from a non-believer stated; "Human arrogance induces the more impressionable into believing we can control climate as if we are a thermostat. We are creating an environmental Tower of Babel where everyone talks nonsense." Our government leaders over the last 8 years placed our economy above the future of the world and refused to take the steps necessary to stop the ever increasing carbon dioxide going into the atmosphere. Facts have very little effect on people who have made up their minds that global warming and climate changes are figments of our imagination. They take the position of, "Don't confuse me with the facts, my mind is made up."

The August 13, 2007 issue of Newsweek magazine had a cover story titled: "Global Warming is a Hoax.* (*Or so claim well-funded naysayers who still reject the overwhelming evidence of climate change.") The article inside the magazine titled: "The Truth about Denial," by Sharon Begley explores the building scientific evidence through the years of the reality that man caused global warming, but even more disconcerting it identifies the organized and well funded efforts to cast doubt that it is occurring. This effort includes some scientists that dispute the fact of global warming, some free-market think tanks, and industrial companies that feel threatened by any action that would reduce the burning of fossil fuels. This includes oil companies such as Exxon Mobil. Their campaign through the years has been quite effective as the Newsweek article points out. In a poll taken in 2006, 64% of Americans thought there was "a lot" of scientific disagreement on climate change; only one third thought global warming was caused by the things people do. A typical position of the naysayers as summarized by Newsweek is, "The global warming over the past century, as well as weather extremes, reflects nothing more than the climate system's normal variability." The consensus view of the scientific world is; "Current warming is 10 times greater than ever before seen in the geologic record. The chance that the warming is natural is less than 10 percent."

However the campaign of denial prevailed within the Bush administration as they consistently set aside any serious effort to stop global warming. Time is ticking away as we continue on our path of adding huge amounts of carbon dioxide to the atmosphere.

An Associated Press newspaper article on July 15, 2008 written y Dina Cappiello titled: **White House ignored EPA report,** reads in part as follows:

WASHINGTON – Government scientists detailed a rising death toll from heat waves, wildfires, disease and smog caused by global warming in an analysis the White House buried so it could avoid regulating greenhouse gases.

In a 149-page document released Monday, the experts laid out for the first time the scientific case for the grave risks that global warming poses to people, and to the food, energy and water on which society depends.

"Risks (to human health, society and the environment) increases with increases in both the rate and magnitude of climate change, " scientists at the Environmental Protection Agency said, Global warming, they wrote, is "unequivocal" and humans are to blame.

Allergies could worsen because of respiratory illness and lung disease, could become more severe in many parts of the country. At the same time, global warming could mean fewer illnesses and deaths due to cold.

"This document inescapably, unmistakably shows that global warming pollution not only threatens human health and welfare, but it is adversely impacting human health and welfare today," said Vickie Patton, deputy general counsel for the Environmental Defense Fund. "What this document demonstrates is that the imperative for action is now."

While the science pointed to a link between pubic health and climate change, the Bush administration has worked to discourage such a connection. To acknowledge one would compel the government to regulate greenhouse gases.

With these kinds of stumbling blocks being placed in the way of any change by the Bush administration there was little hope of any serious effort being made by the United States to solve the problem of global warming until Bush was out of office.

Australia is one of the countries along with the United States that have refused to sign the Kyoto Protocol. It is ironic because Australia is one the areas of the world that is already suffering from significant climate change. Large areas of the country have been experiencing severe drought. Perth on the West Coast of Australia faces the loss of their fresh drinking water as they draw down the water reservoir that normally has a four year supply but now could be gone in less than a year. However, a change of leadership in Australia in early 2008 has brought a changing view of global warming in Australia.

The Southeast part of the United States has also been suffering drought conditions that have resulted in serious wildfires in Florida and Georgia. One of the largest tornadoes ever recorded totally destroyed Greensburg, Kansas in the spring of 2007. In the first half of 2008 the tornado incident rate was particularly high.

• • • • • • • • • • • •

In the October 2007 issue of National Geographic Magazine there is an insert that dramatically illustrates what has happened on our planet over the last 400,000 years. It also shows what will be happening in the near future if we continue on our current path. It clearly plots the level of carbon dioxide in the atmosphere, along with the change in temperature both above and below the current level. Sea level changes above and below current levels are also shown. This set of data should absolutely dispel any arguments that say that the carbon dioxide level in the atmosphere does not cause global warming. It is also absolutely clear that the current dramatic rise in the level of carbon dioxide is the result of human activity in burning hydrocarbon fuels. The correlation between carbon dioxide level changes with temperature and sea level changes is amazing. Over the last 400,000 years prior to the start of the industrial revolution the level of carbon dioxide in the atmosphere has varied through natural phenomenon over a range of from about 180 parts per million to a high of about 300 parts per million. The pre-industrial revolution level was 280 parts per million.

Today the level is 380 parts per million. The most frightening aspect is the dramatic increase in the level of carbon dioxide in the next few decades if we continue to burn hydrocarbon fuels at our current rate, the level will hit 525 parts per million by 2100. As National Geographic says, "Business as usual could push CO_2 levels above 800 ppm, triggering temperature hikes of up to 9°F and likely overwhelming the ability of many species to adapt."

This will have a devastating effect on planet earth and all the species that inhabit it.

In a speech about energy and the future on July 18, 2008, Al Gore said, "The survival of the United States of America as we know it is at risk, the future of the human race is at stake." As David Stout reported in The New York Times, Gore called for the kind of concerted national effort that enabled Americans to walk on the moon 39 years ago, just eight years after President John F. Kennedy famously embraced that goal. He said the goal of producing all of the nation's electricity from "renewable energy and truly clean carbon-free sources" within 10 years is not some farfetched vision, although he said it would require fundamental changes in political thinking and personal expectations.

• • • • • • • • • • • •

We know that our current highly evolved lifestyle depends on energy obtained by burning stored fossil fuels that are the main source of excessive carbon dioxide in the atmosphere. We have to stop this and develop energy sources that do not contribute carbon dioxide to the atmosphere. If we don't, the consequences will be very unpleasant. The world may not cease to exist, but it will no longer be the world we know.

We must replace oil and coal as our primary energy source. Unfortunately we have waited too long to start, so we need to act as rapidly as possible to reduce our consumption of oil and coal, while we develop a replacement. As you have seen in the last chapter there are a number of renewable energy sources that we can implement quickly to reduce the level of carbon dioxide introduction. The quickest way is conservation and increased energy efficiency. If we doubled the fuel mileage requirements of our cars and light trucks the effect would be dramatic. It is well within our capability to reach the level that Japan and Europe have already achieved. The reduction in carbon dioxide could be significant since we use 25% of the world's oil and a great deal of it is in personal transportation.

This is probably the largest single step we can do in the short term, but also one that has tremendous economic and personal implications. It means scrapping most of our existing personal vehicles and replacing them with smaller high mileage vehicles. It is unfortunate that we have had leaders that were blind to what was going on around them and failed to face up to unpleasant choices. If they had moved years ago to start the process the world would be reaping the benefits today. Now it is going to be hard to convince all the moms driving the kids to school to give up their big SUVs and guys won't want to give up their big pickups. Maybe it can be like what is happening on our island. The in-thing might become the little electric cars and trucks we are seeing on the roads.

If we as a nation had the foresight to apply a significant fuel tax of one to two dollars a gallon or more years ago we would already have changed to high fuel economy vehicles as most of the rest of the developed world has already done. In addition the revenue generated would have given us the funds to develop solar power satellites years ago. The simple factor of cost would have forced us to be driving high mileage cars instead of the gas guzzlers that clog our highways today. The new CEO of Ford proposed that adding a significant fuel tax would be the best way to force the consumer to buy high mileage vehicles, rather than having Congress force the manufactures to increase mileage, which consumers currently do not want. As the price of gas moved above $4.00 a gallon the transition to higher mileage vehicles became evident. Sales of big pickups and SUVs has plummeted. Orders for Toyota Prius Hybrids were eight months ahead of the supply and American companies are in a panic to expand their lines of high mileage vehicle. So the evidence is clear, if we had placed a high tax on gas and diesel years ago we would not be consuming 25% of the world's oil usage today.

The transition we face today would have been so much easier and the peak oil event would be farther in the future if we had switched to high mileage vehicles years ago. As we change our personal vehicles we are at a time of technology development that gives us some options. These include small gas powered cars, high efficiency diesels, and hybrids such the Toyota Prius. Plug-in hybrids are on the horizon with very high mileage capability. A potential variation on these is solar cell covered vehicles that use solar energy to charge the batteries. Pure battery power can be used for short range driving using our existing battery systems. Battery technology is moving rapidly as I will discuss in more detail in a later chapter.

The transition to advance battery powered cars has already started with General Motor's introduction of the Chevrolet Volt The February 19, 2007 issue of **People** magazine carried a brief article on actor George Clooney's plan to tour Italy in his lithium ion battery powered Smart Car. It gets 120 to 150 miles on a single charge costing $2.10. There are more options that will become available as we change from an oil based economy to an electric based economy, but this is a precursor of the future.

The good news in this traumatic change is the opportunities it will present to those companies who understand what must happen and are there to supply the need. It is extremely doubtful that all of today's automobile manufacturers will be able to make the transition. Like our leaders they have been more focused on short term profit. Our political leaders seem to be more interested in staying in office than the long range future of the world. The environment be damned, they won't be around to worry about it.

After the 1973-74 oil embargo we saw a significant drop in US oil consumption that resumed its growth only after we thought the energy crisis was over. Unfortunately, we did not have a good understanding of global warming at that time, or we may not have been so casual about going back to our old ways.

There are countless other ways to decrease our use of energy and all of them should be used. If we are careful in applying them we do not have to greatly diminish our standard of living. To illustrate this I will tell you a story of our experience living a life of conservation.

After retirement from Boeing in 1987, my wife and I set off from Seattle in our 49 foot sailboat with no specific goal in mind except to be in Cabo San Lucas by Christmas, and then to let the winds of chance, lead us on. We ended up spending six years cruising in Mexico and the South Pacific. During that time we only took the boat to a dock when we put on fuel, which was about twice a year, or to haul out for bottom painting every two years. Other than that we were totally dependent on ourselves and the boat to supply our basic needs. We used our dinghy to go ashore and bought food at the local markets and caught fish when at sea. We loaded water at some locations where we were sure it was pure, but most of our water came from rain catchments during the many squalls in the South Pacific. We did have a small water maker that could make 1.3 gal-lons an hour of fresh water from sea water. For energy we had a wind generator and a large alternator on our main engine to charge batteries and a small auxiliary diesel generator that supplied both 110 volt AC current and 12 volt DC to charge batteries. In addition we had a small inverter that ran off the batteries that we could use to power a TV and video player. I wished that I had installed solar panels like many of the cruising boats used, but our wind generator did a good job for us. Living for six years with the forced limitations imposed by limited water and power and access to supplies taught us how comfortable we could really be without all of the excesses we were used to in our home ashore. In a world of exploding population and limited resources we need to adopt the concept of sustainable living if we are to leave a world for our children and their off spring.

If we look around our homes we can find numerous areas where we can conserve energy. It will not only help the environment but it will also reduce our costs of living. The development and installation of alternative renewable energy sources that can be applied quickly is another avenue that needs to be aggressively pursued. As we move into the era of high energy costs and a collapsing economy we will be forced to make reductions in life style because that is what we can afford.

The contribution coming from wind turbines is already significant in Europe and we need to grow it in the

United States. The development of middle America was heavily dependent on windmills to pump the water to sustain both animals and their owners, so wind power is not new to the United States, and its now time to reinvigorate this source of energy. There are many areas around the country that have sufficient wind to make wind power viable. Even on our little island our co-op has set up a tower to measure the wind profile in order to identify suitable sites for small wind turbine installations. Today there is a development of homes being built that will use wind and solar for most of their power needs.

Terrestrial solar power needs to be expanded greatly. Fortunately, today new technologies are increasing efficiency and reducing costs. Even in the northern parts of the country they can make a significant contribution. Some of the solar systems now being manufactured are designed as roofing material with imbedded solar cells for generating power. This type of installation can be integrated into new construction or as roof replacement. They are attractive and blend into the rest of the home design. There are also windows of various colors and design that are energy producers. As development progresses all homes could eventually be providing much of their own energy. At least they will be able to supplement their requirements.

Our energy needs in a modern society are of such a high magnitude we need to develop all the systems possible in order to reduce our use of hydrocarbon fuels as quickly as possible. We can't stop all of the effects of global warming because of all the carbon dioxide already in the atmosphere, but we can slow its increase while we work to stop burning fossil fuels.

This will buy us time while we develop and build solar power satellites as the replacement for fossil fuels. It is an absolute requirement that we stop burning fossil fuels if we are to stop global warming and stopping global warming is essential to the survival of our world as we know it. At the same time if we are to continue to support our civilized world's population and provide for the development of the emerging nations it is also absolutely essential that we provide the energy to make this possible. The clear solution is the development and deployment of Space Based Solar Power.

One day several years ago after we returned from cruising the South Pacific, I was working on refitting our sailboat while it was hauled out at a marina. I had stopped to take a break and have a beer, when my cell phone rang. A voice from the past was on the phone. It was an old friend from NASA. I had worked with him when I was part of the Grumman/Boeing Space Shuttle definition study team and later on during the Solar Power Satellite Program. His first words were, "Ralph I called to apologize. I have felt bad for years and I just had to call and tell you." I didn't know what he was referring to and asked him what he was apologizing about. "I was the one who was assigned to kill your design for the reusable booster for Space Shuttle, it was a terrible mistake, if we had gone ahead with your configuration the world would be totally different today," was his answer. He went on to suggest we jointly write a book about what the world would have been like if NASA had selected the reusable fly-back booster for the Shuttle instead of the troubled configuration we are all familiar with today. We did not go ahead with the book idea, but I am going to explore the possibility in this chapter. In order for you to understand what happened I will give you the background first.

When it started I was based at the Michoud plant in New Orleans, Louisiana where Boeing was building the Saturn S-IC which was the first stage of the Saturn V moon rocket. It was the largest rocket ever built. I had been the one that developed Boeing's proposed configuration that won the contract to design and build the first stage and then became the manager responsible for the structural design of the fuel tanks. I was later assigned as Cost Effectiveness Manager for Saturn V as we were working to develop follow-on versions for other missions after the moon program. We were also studying an alternate booster concept. At the same time NASA was initiating studies of a fully reusable vehicle they called Space Shuttle. The Boeing Space Division in Kent, Washington was gearing up to bid on one of the Phase B Definition Studies, but Boeing management realized that all of the experienced rocket people were at Michoud. As a result several of us received telephone calls telling us we were to report back to Seattle immediately. When we arrived it was to a chaotic situation. Boeing was in the midst of huge layoffs as the airline industry was in one of its cyclic downturns, the Supersonic Transport Program had just been cancelled, and the company was having serious problems. You may remember articles about the billboard sign near the airport that some of the laid-off workers had arranged for that read, "Would the Last One Out of Seattle, Please Turn Out the Lights."

The manager who was in charge of the bid was not one of Boeing's best. He was more interested in empire building than winning the contract. As a result he had assembled a team that was much too large for the program and they had practically no launch vehicle experience. In addition they were nearly all the lower rated engineers who were in line to be laid-off. He had also signed up with Lockheed to be our partner for the orbiter part of the study. Lockheed at that time had been one of the Phase A contractors, but did not have any significant space transportation experience and did not have a good reputation in the industry. The Michoud contingency took about a week to get oriented and assigned to critical jobs. My role was to have responsibility for booster structures design. However, before we could really start on the task, the first job before us was to notify about two thirds of our team that they were being laid-off. It was one of the hardest things I have ever had to do.

The study we were bidding on was what NASA called a Phase B Space Shuttle Definition Study. Nearly all of the major aerospace companies in the country were paired off in teams to make a bid. NASA planned to award multiple contracts so there would be continued competition during the studies. The criteria established by NASA was for a two stage fully reusable fly-back system with both stages powered with liquid hydrogen and liquid oxygen, with a maximum lift-off gross weight of three and a half million

pounds. The payload bay was to be 15 feet in diameter and 60 feet long with a payload capability of 65,000 pounds. It soon became apparent we had a very high mountain to climb. Lockheed was proposing a delta shaped lifting body for the second stage which was the orbiter stage of the system. Our part of the job was to design the booster stage.

We worked up several booster concepts with wings and jet engines to provide for fly-back to the launch base, but they all turned out to be too heavy to meet the three and a half million pound gross lift-off requirement. The low density of liquid hydrogen demanded a large tank that made the vehicle quite large. The added problem was that the center of gravity during launch with propellants aboard changed for the fly-back part of the mission. We were banging our heads against the weight wall and not making much headway. The deadline for submitting the proposal was approaching and we still didn't have a configuration that would meet the NASA criteria.

One day a group of us from Boeing were at lunch with some of the guys from Lockheed in the cafeteria at the Lockheed facility in California brainstorming ideas when a thought struck me. Why not think of the booster as an airliner that carried its fuel in the wing? We couldn't carry the hydrogen, it had too much volume, but we could carry the oxygen. That would also solve our center of gravity problem. I sketched out the configuration on a napkin and it turned into our proposed configuration. It met the weight target and was a really pretty vehicle. The wing design was a little like an air mattress with tension ties between the top and bottom surfaces, so it could take the pressure required in the liquid oxygen tank.

We submitted our proposal along with all of the other contractor teams. We waited with baited breath during the NASA evaluation of the proposals. When it came we found out we had lost. However, immediately following the notification that we had not won one of the multiple Phase B contracts, the Director of NASA called our Space Division Vice President and asked if Boeing would be willing to team with Grumman and work as the booster contractor with them on a study of alternate configurations for the Space Shuttle.

During the period that all the aerospace companies were forming their two company teams and preparing to bid the contract, Grumman found themselves the odd man out and did not have a booster partner. They had studied the concept and concluded that NASA had their head in the sand on the weight target. Rather than try to fight a battle they knew they couldn't win they stepped back and asked the question , "Is there a better solution?" They came up with one and decided to submit a non-responsive, but very innovative proposal to NASA. Their proposal was that using hydrogen as the fuel for the first stage was the wrong answer and it should be a

hydrocarbon fuel and that a gross lift off weight limit was not a good criterion. They proposed that they conduct an alternate configurations study that considered other systems besides hydrogen for the first stage. They also proposed that a logical start was using the Boeing S-IC with wings as the booster. The S-IC was the first stage of the Saturn V that was used to launch Apollo to the moon for the manned lunar landings. The S-IC had seven and a half million pounds of thrust to lift-off. The Director of NASA liked the idea and since Boeing was now available, and the builder of the S-IC, we were back in business. A week later I was on my way, along with the rest of our team to New York that was to be my part time home for the next two years on Long Island, Grumman's home base.

Many months later when I was briefing the Chief Engineer of NASA's Manned Space Center in Houston (now the Johnson Space Flight Center) on our progress, he stopped me and asked, "how on earth did you ever come up with the idea of putting oxygen in the wing of your booster, it could never work." Then I knew why we had lost the first contract bid. Through the following years he and I had many arguments, and I convinced him some of the time, but he had very fixed ideas and was reluctant to change his mind. Many years later, when I was visiting the Smithsonian Air and Space Museum in Washington DC, there on display was a model from our original lost proposal of our booster configuration that carried the oxygen in the wing. I have no doubt that some time in the future when we build reusable launch vehicles and stop using hydrocarbon fuel as our first stage propellant there will be a hydrogen booster with oxygen in the wing. I don't know why the Smithsonian had selected that vehicle to display, except it was the prettiest one in the competition.

After we became Grumman's partner Lockheed protested that we had been given a contract without a competition, so to placate them NASA gave them a contract also. That meant that now all of the large aerospace contractors in the country were working on Space Shuttle Definition contracts.

The Grumman/Boeing contract was different from the rest, and NASA called it a Phase A contract because we were to evaluate many different configurations while the other contractors were concentrating on the NASA specified concept using hydrogen fuel in the first stage. Now that the Chief Engineer at the Manned Space Flight Center was involved he insisted that there should be two approaches to the wing configuration of the hydrogen fueled booster. One of them had to be a straight wing. He had been the inventor of the blunt ballistic reentry shape of the Mercury, Gemini, and Apollo capsules. His theory on the booster shape was that you simply cut out an airplane shape from the circle formed by the blunt capsule and have the vehicle reenter the atmosphere in the same manner as an Apollo capsule. All the contractors were force to consider this approach as an option.

Our first task was to establish what kinds of configurations we were going to develop and compare. This would include two stage systems using solid rocket boosters and stage and a half configurations using solid rocket boosters. Then there was one we called "big dumb booster." We called it dumb because it was a very simple design, low-cost, throw-away booster using pressure fed engines with no pumps. Also included in the group were hydrocarbon fuel boosters that were recovered ballistically with parachutes. In addition we studied NASA's reference configuration which was a hydrogen fueled fly back booster, even though they would not let us include oxygen carried in the wings. Our design of choice was always the S-IC converted to a fly-back booster with a delta wing. I was assigned as the manager of this configuration.

When Boeing became Grumman's partner, management realized they needed a new leader for the effort that understood airplanes as so much of the effort would involve aerodynamic flight. As a result Boeing assigned a Vice President from the airplane side of Boeing to be Program Manager who was well respected in the aerospace industry.

My involvement with the big dumb booster started with an impulsive comment I probably shouldn't have made. It went back to the period I was cost effectiveness manager for the Saturn V at Michoud. Our Space Division Vice President was in New Orleans for his monthly program review which I sat in on. He was telling us about a booster program that was being studied by Boeing in Seattle that was a low-cost pressure fed booster. He was all excited about the tough steel called HY-140 that was going to be used to make the propellant tanks. He said the welds were so strong they couldn't be broken. That remark got to me as I had spent my career at Boeing working on welded pressure vessels. I jumped up and interrupted him and said "George, you are wrong; any welds can fail if they have defects." There was absolute silence in the room as we stood and stared at each other. I figured I was in real trouble. Finally he said, "I am going to find out who is right," and went on with the review and I sat down.

When he came back for the next monthly review, he started the session by saying he had looked into the situation on welding HY-140 steel and determined I was right. He went on to say, "I am transferring the program to Michoud and Ralph, you are to take over as Design Manager."

I don't remember when we started calling the vehicle big dumb booster, but that is how it became known. It was an interesting concept and we worked to make it even simpler and more efficient. Since the tanks had to be pressurized to a high pressure to feed the engines we looked at optimizing the shape of the tanks to be more efficient. We started with cylindrical tanks because that is what most rockets used,

but we knew that spheres were more efficient pressure vessels. This led to a change in the design that used two nested spheres to carry the propellants. This made it look like a two dip ice cream cone. The next change was to eliminate the high pressure helium tanks that were used to pressurize the propellant tanks. We could do this because the propellant for the system was hypergolic. Hypergolic meant that the propellants would ignite instantly on contact. That allowed us to design a simple system that squirted a small amount of fuel directly into the oxidizer tank and a small amount of oxidizer into the fuel tank. This resulted in controlled ignition of the propellants inside the tanks and pressurizing them with the hot combustion gases. We even built a small scale vehicle to test the concept. One of the problems with the concept was that the fuel and oxidizer was pretty nasty stuff and the engine exhaust contained a lot of pollutants. However, in those days we didn't worry too much about atmospheric pollution. We were still working on the concept when we were called back to Seattle to work on the Space Shuttle Definition studies.

One of the stage and a half configurations using solid boosters and a huge expendable tank to carry the propellants for a hydrogen powered orbiter ultimately became the Space Shuttle we have today. At our first major review there were about 20 configurations which we evaluated and made recommendations to NASA as to their merits. The evaluation criteria included development costs and operations cost estimates. As part of our recommendations we told NASA we should drop the stage and a half configuration along with several others. We told them that in addition to its operations cost it had several other inherent problems. The dynamics of the system would be difficult and it had no growth potential. They accepted most of our recommendations, but told us to keep the stage and a half. They also told us to drop the big dumb booster. It was a little too elegant in its simplicity for NASA to understand, even though we had included parachute recovery in the ocean as part of the study. It was robust enough that it could take the reentry and water landing without any significant problems.

This cycle of deeper studies and reviews continued. Each time we came to present our findings and we recommended eliminating the stage and a half and they told us to keep it. At the same time, the fly-back S-IC configuration looked better and better. The studies of the other contractors working on the hydrogen fueled boosters were not making much headway in reaching the weight target. Also, since Lockheed did not have a booster contractor working with them, we were required to submit all of our booster data to Lockheed to support their work. This was a bizarre situation. I was called on to do most of the briefings to the Lockheed team. It was always a strange feeling to walk into their facility. They had started as our team mates and now they were our competitors and I had to tell them everything we were doing on the boosters.

NASA finally came to the conclusion that the cost and weight of the all hydrogen system wasn't going to work at their target weight and the costs were too high. They decided that they would modify all the contracts to work on the configurations we had developed. The Grumman/Boeing contract was upgraded to a full Phase B status and we were directed to brief all of the other contractors on our configurations. We were now down to two basic versions besides the NASA baseline of all hydrogen, the stage and a half solid booster with the huge external tank and the fly-back S-IC that we called the RS-IC. The Grumman designed orbiter for this configuration used five of the J-2 hydrogen engines that had been used on the second and third stage of the Saturn V moon rocket. It also had small external hydrogen tanks that were mounted on each side of the orbiter payload bay. These tanks needed to carry only a fraction of the hydrogen required for the stage and a half system. The reason was because in a two stage system the booster stage does the majority of the heavy work and the second stage propellant requirement is relatively less. In the case of a stage and a half system the final stage has to burn all the way from the ground to orbit so its fuel requirement is very large.

Since I was the configuration manager for the fly-back RS-IC booster it was my job to brief the other contractors. It was a strange feeling to walk into their midst and tell them about what they were going to have to do. Hostility permeated the atmosphere in the rooms. How could this have happened? They had won their contracts and we had lost and here I was telling them about a vehicle that we built that they had to incorporate into their design. They were also briefed on the stage and a half vehicle which had also been developed by the Grumman/Boeing team.

As we moved towards the final contract reviews, recommendations and configuration selection by NASA, there was a great feeling of uncertainty about what was going to happen. Nixon had been elected President, the greatly respected Director of NASA had retired and his replacement was a bureaucratic politician. The future status of the main NASA centers at Huntsville, Alabama and Houston, Texas was uncertain. The chairman of the Senate Space Committee was lobbying for business for his state of Utah. There was no feeling of commitment to a common goal as there had been for the moon program. In this atmosphere it was every one fighting for their own little chunk of the future.

To give you a little more background on the situation concerning our proposal on the fly-back S-IC it is important to tell you that we had two flight ready S-IC stages in bonded storage at Michoud that were left over from the lunar landing program after Nixon had cut it off. We proposed modifying these two vehicles by adding wings, tail, and jet engines in an extended nose fairing. We had extensive wind tunnel data that verified the flight configuration and the operation of the jet engines in the nose extension. We also had the moon launch experience to verify the launch performance. It was an all aluminum vehicle and because of the thickness of the wing skins did not require any additional thermal protection for re-entry. It is what we called a heat sink booster. The maximum re-entry heat would not raise the temperature of the material over 350 degrees Fahrenheit which it could withstand for repeated cycles. We had enough spare F-I engines to fly more than 400 missions with refurbishment of the turbo pumps. The F-I was designed as an expendable engine, but was certified for a sufficiently long duration of firings to fly many missions before it had to be refurbished. In addition, the engine design had been up-graded from 1.5 million pounds of thrust to be able it to produce 1.8 million pounds with a new turbo pump that had also been certified, so we had growth potential. The Grumman designed orbiter used the J-2 engines that had been used on the second and third stage of the Saturn V, so no new engine development was required. In contrast to the stage and a half configuration which required development of a new ultra-high efficiency, throttlable, hydrogen engine, development of new solid boosters in addition to the orbiter and huge external tank.

The logical configuration selection was the one based on the modified S-IC with the orbiter using small hydrogen tanks and the J-2 engines from the Saturn program. By this time we were estimating the total development costs at about the same level as the stage and a half system, and the operations cost only a fraction of the stage and a half. The program risk would be low because so much of the hardware was already proven by the Saturn program. As we were to find out this was a recipe for disaster in the political environment that existed in Washington DC and the NASA centers at Huntsville and Houston.

Let me explain the situation in Huntsville (The Marshall Space Flight Center). They were basically a launch vehicle and engine development center. If we were to go forward with the fly-back S-IC configuration, there would be no engine development or booster development required. Since Boeing was a builder of large airplanes they certainly wouldn't need Huntsville's help to convert the S-IC into an airplane. On the other hand the stage and a half concept required the development of an exotic new hydrogen fueled engine and the solid rocket boosters and the external tank. Their responsibility would be to oversee the development. This configuration assured their future.

On the other hand, Houston was the Manned Space Flight Center (now the Johnson Space Center). They would be in charge of the orbiter development for either configuration, but they had one big fear. They knew we could modify the S-IC in two to three years and have it ready to fly while it would take them at least seven years to develop the orbiter. Their fear was that if we had a booster ready to go we wouldn't just let it sit there idle. We would fly it with some other expendable upper stage. With the administration in Washington DC in the mood to save development money, they would never get to finish the orbiter.

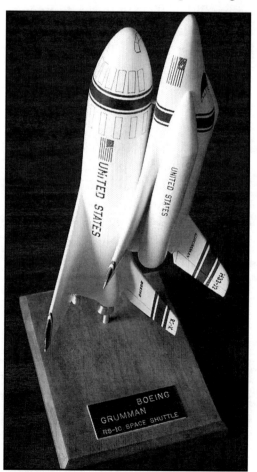

Figure 2: Model of the Boeing fly-back RS-IC booster (Saturn first stage S-IC with wings, tail and jet engines) and the Grumman H 33 orbiter using J-2 engines

In Washington DC we had two problems. First the administration wanted to keep short term spending to a minimum and didn't really care about how that would effect later funding. Second, the Senator from Utah who controlled the Senate Space Committee and was very influential with the new NASA administrator wanted money to go to the solid rocket manufacturer in his state. A supposedly independent cost analysis was conducted comparing the various configurations being considered. When it came out we knew we were in trouble. It grossly changed the cost of the concepts from what we had developed to warp the results towards the stage and a half.

The day finally came for all the contractors to present their final reports and recommendations in Washington DC before the new Director of NASA, the Center Directors, all of their senior managers, and study managers. All the contactor teams were to present on the same day at different times. I was selected to present the Boeing part of the Grumman/Boeing briefing. In addition I was to sit in on the Lockheed briefing as they did not have a booster contractor. The ground rules that NASA established for this strange situation were that I was not to say a word, unless I was asked a question by a NASA member, and then only to answer the specific question. Lockheed was to go on first and the Grumman/Boeing team would be the third ones to present.

After the change in the configuration emphasis had taken place, Lockheed's focus had changed. They had no hope of being the contractor to modify the S-IC so they had switched their emphasis to the solid booster configuration as they were trying to expand their business base with solids.

When I walked into the briefing room the Director of NASA introduced me and explained to the audience why I was there in the Lockheed briefing. I knew most of the Lockheed people from the brief period we had been partners and from the booster briefings I had given them. It was not long into their briefing before the anger started building in me from what I was hearing. By the time it was over I was nearly shaking with fury. It was good that I had not been asked any questions, I don't think I could have controlled myself to just answer. Their presentation was totally biased and distorted the facts we had given them so that they weren't even recognizable. It confirmed why they did not have a good reputation in the industry. I was glad we were no longer their partners. It made me think back to another incident that had occurred when we were working with them on the early contract bid. It was late at night and I was on my way out of our facility when I passed by the Lockheed area and noticed their chief of aerodynamics working away on a graph. I stopped to ask him what he was doing. He said he was working on the aerodynamics chart for their orbiter for the proposal. "But, I didn't think you had the wind tunnel data yet" I said. "Oh, we don't, but this is what it has to look like if it is going to work," was his reply. I should have known what to expect from them.

By the time I got to the Grumman/Boeing area I had calmed down and told them what I had just heard. But, the downhill slide of the day was nowhere near over. What had happened while I was in the Lockheed briefing was the Grumman folks decided they needed to escalate the level of the presenters. Instead of the Grumman Program Manager giving the majority of their part of the briefing they assigned it to his boss who was a Vice President, and escalated the two other parts of the presentation to Senior Vice Presidents. None of them had been prepared with the changes and we were due up in less than half an hour.

We all walked into the briefing room and found our seats. The Senior V.P. from Grumman started the briefing as they were the leaders of the partnership. I knew we were in trouble with his first words as he stumbled in his introduction. It had been planned that Grumman would open with a strong statement of what we were proposing which was the fly-back S-IC with Grumman's orbiter using the J-2 engines and saddle hydrogen tanks. It came out as a wishy-washy statement with no impact. The next presenter did a little better and then it was time for the body of their presentation which when given by their Program Manager during the dry run was great, but the

V.P. that got up was obviously extremely nervous. He had a mike in his hand with a long cord that led back to a plug in the wall under the projector screen. This was in the days before cordless mikes.

As he started to talk he was so nervous that he started winding the mike cord around his arm without realizing what he was doing. Everyone in the room was mesmerize by his actions and didn't hear a word he said. He literally wound himself up to the wall and suddenly stopped with a confused look on his face as he realized what he had done. To break the tension the Director of NASA stopped him to ask a stupid little question to give him a chance to compose himself. The break was sufficient to get him back on track and the remainder of his briefing went fairly well but the damage was done.

I was up next. I knew the obstacles we faced were nearly insurmountable, but I gave it my best shot. I expanded my briefing to try and fill some of the holes left by the Grumman presenters and to counter what Lockheed had told this same audience that morning. In addition we had to overcome the distorted cost comparison study and the bias of the NASA Centers. It was not enough.

I had just gone through the worst day of my life. I believe it was one of NASA's darkest days also, as they selected the stage and a half configuration with the solid rocket boosters. The United States, the world, and 14 dead astronauts have paid the price ever since.

Back to my friend's apology. His bosses at Houston had given him the task of choreographing the information stream and studies, including the supposedly independent cost analysis, that would show that the fly-back S-IC was the wrong choice. He was obviously successful, even though it had been a terrible mistake.

Today the two flight quality S-ICs are museum pieces at Cape Canaveral and the Johnson Space Center in Houston.

• • • • • • • • • • • •

In 1981 I represented Boeing at the first launch of the Space Shuttle. It had taken 9 years to develop the system from that disastrous day when they selected the solid rocket booster/external tank configuration. The development costs had overrun the NASA projected costs dramatically. Also during that time I had gone through two more traumatic career setbacks.

The first occurred in 1973. After we failed to convince NASA to use the fly-back S-IC as the launch vehicle for Space Shuttle, Boeing elected to bid on the External Tank contract for the chosen configuration. It was to be built at Michoud in New Orleans where we had built the S-IC. The bid team was formed around the group that had worked the Shuttle Definition Study in Seattle, but we were soon transferred

back to New Orleans. The design concept for the tank was very similar to the structure we used on the S-IC propellant tanks and the material was to be the same. We had a good grasp of the design and what it would cost. One of the differences was the need to apply foam insulation for the liquid hydrogen tank. In addition it was necessary to insulate the liquid oxygen tank because its location forward of the orbiter could cause ice falloff during launch that could damage the orbiter. The oxygen tanks on boosters were not normally insulated as ice falling off during launch was not a problem without something attached at the side.

Boeing selected as Program Manager a former NASA employee who did not have any hardware experience. As the bid process proceeded he panicked and asked Seattle for help. After our greatly respected Space Division leader recently died, his replacement decided instead of helping the program manager to replace him with a former small satellite manager who had no launch vehicle experience. I was also replaced as chief engineer with a Seattle transplant and demoted to Design Manager. Our new Program Manager soon had our customer pulling their hair wondering what was going on. I knew we were in big trouble.

One feature of our design that could have had a major impact on the success of the current Space Shuttle was the method we developed for applying the exterior foam on the tank. Instead of simply spraying the foam on the tank our installation process used a moving mold. The foam was introduced under the mold as the tank was rotated under it. The moving mold produced a smooth hard skin coat over the foam. This technique may have prevented the foam loss in flight that has been a chronic problem from the first flight and did cause the loss of Columbia.

NASA had established a cost per tank target for the bidders to meet. We had very hard data on what it would cost to build the tanks because they were so similar to the S-IC fuel tanks we had been building for the previous ten years in the Michoud facility The material was the same 2219 aluminum alloy and the design concept was the same. We knew we could not meet the target, but we could come close. Our competition included Martin Marietta who was building the Titan. We all submitted our bids and the winner was Martin who underbid us by about a million dollars a tank. However, following typical government contract bidder's policy of bidding low and then making it up with change orders, Martin didn't even have to do that. They were able to get the per tank cost increased to nearly four times their bid before the contract was even signed. Boeing shut down their operation at Michoud and turned the plant over to Martin.

That was not the end of the cost increases. Several years later, NASA came to us and asked if we would be willing to re-bid the contract, since the Martin cost had gone so high. We agreed to evaluate the potential. NASA made Martin open the Michoud facility and their manufacturing cost records to us. We were able to show NASA that Martin's cost was five times what Boeing's costs were to build the S-IC tanks. We did it in man hours per pound of hardware so there were no financial gymnastics involved. Instead of re-bidding the job NASA said, "thank you," and used the data to force Martin into a contract cost reduction, but it was nowhere near the cost we could have done the job for.

The second major setback came in 1980 with the termination of the government's solar power satellite program. But, first I want to explore "what if" we had the fly-back system for Space Shuttle instead of the system we now have.

As I watched that first Shuttle flight in 1981 it brought back the memories of watching Saturn V lift off for the moon on man's greatest exploration. We were going back to space for the first time in many years. Those of us at the launch soon started hearing murmurs that there were concerns from the launch crew of debris falling off the external tank that might have damaged the orbiter during lift-off. It was some time later that the word spread that it was OK. Apparently, the Air Force had a satellite that could make a photographic scan of the orbiter in orbit that showed no visible damage and the flight was successful.

NASA's original plan had called for at least 400 Shuttle flights, but it was not to be. The turnaround time between flights was much longer than originally planned, and then the Challenger accident happened when an O-ring on a solid rocket booster failed catastrophically, killing 7 Astronauts. The program was grounded.

The weather had been particularly cold for some time before the launch, but there was a lot of pressure to go ahead with the launch because of all the delays the Shuttle program had experienced through the years. So the launch was approved even though the temperature was below NASA guidelines.

As the investigation of the cause of the accident proceeded the details of the O-ring design and installation were revealed. NASA had required a pressure test of the O-ring seal to verify that it sealed the joint. The only problem with the test they specified was that it was on the wrong side of the O-ring and so it unseated the ring from the side that took the operational pressure. This was not normally a problem because O-rings by their very nature move in their grooves, but in the case of the Space Shuttle disaster, the cold weather had hardened the material to the point it could not reestablish a seal when exposed to the positive pressure of the launch firing. NASA never admitted that their test was the basic problem. The cold weather launch was blamed. After a prolonged three year

period of rework and finger pointing the program was resumed.

• • • • • • • • • • • •

The loss of foam from the external tank continued, but nothing significant was done about it as NASA did not consider it to be a major problem. Then came the disaster of Columbia due to ice and foam impact on the wing that caused its disintegration on re-entry and 7 more Astronaut lives were lost. It is flying again, but with plans to retire it in 2010. NASA has spent bundles of money trying to solve the foam problem on the external tank but have failed to make the one change that could really fix the problem. They need to wrap the foam in the area ahead of the orbiter wing with some material that will positively retain the foam in place. This needs to include the capability to attach local wraps over the fittings. Why they have failed to make this simple fix baffles me.

In addition, weight growth over the years on the system had reduced the desired payload and NASA finally came to realize we had been right and that the configuration did not have growth capability. Reduced payload capability has plagued the program since its beginning.

The per flight cost of flying Space Shuttle has absorbed huge amounts of NASA's budget and dramatically reduced their ability to perform other missions. In addition the space station assembly has been stalled for years and is just now getting back on track.

The Shuttle was supposed to be reusable and the orbiter is, but the throw away external tank which was supposed to cost one and a half million dollars each, now costs many, many times more. The money spent on trying to fix the foam problem has undoubtedly driven the cost higher still. The solid rockets are recovered and reused, but their refurbishment and cost of the fuel alone makes them very expensive. The decision to select this configuration was justified by its projected low development cost. In reality because of the high cost of developing a new hydrogen engine, the solid rocket boosters and the external tank, and the orbiter, its actual cost was much greater than the cost would have been to modify the S-IC and develop just the orbiter. It took nine years. The S-IC could have been modified in less than three years and the orbiter alone without the need for a new engine could have been 6 to 7 years. In government contracts, time is one of the most costly elements.

As we look back at the cost of operating the Space Shuttle which was supposed to dramatically reduce the cost of launching payload to orbit, we find it has the highest cost of all systems. It should have flown at least 400 times by now instead of a little over a 100 fights. Its turnaround time that was supposed to be weeks is instead months. The giant external tank is thrown away on each flight. The solid rocket boosters are expensive to refurbish and recast their expensive propellant. If the Shuttle had used proven propulsion of the fly-back S-IC booster the turnaround time would have been reduced, the cost of propellant would have been much less. The small size of the orbiter hydrogen tanks would have cost much less. The orbiter mounting on the back of the fly-back S-IC would have prevented the possibility of ice or foam damage to the orbiter. The overall result would have been dramatically reduced costs.

In addition to the lower launch cost, the other great advantage of the S-IC was its inherent growth potential. The F-I engine had already been certified to 1.8 million pounds of thrust with a redesigned turbo pump from the original 1.5 million pounds. The fuel capacity could have been expanded with propellant tank inserts. This would have made the construction of the space station much simpler and quicker.

Enough comparisons. What if the fly-back booster system had been developed? Its first flight would have been in 1978 or1979. What if that had happened during the time we were under contract to define the solar power satellite system? We would have had the prototype for the launch system to show the Department of Energy. It would not have been large enough for full scale satellites, but it would have been solid evidence that the launch system we were proposing would work. What if we had shown the people of the world that access to space could be routine and dependable?

The space station would have been completed and fully manned by now. It could have started to do many of the experiments and demonstrations that would have been a great help for the solar power satellite program and many others.

While the United States struggled with problems with the Space Shuttle, Russia has filled the need to fly supplies and crews to the International Space Station. They even fly commercial paying passengers, first to the MIR station and later to the International Space Station using their Soyuz vehicle. If we had the fly-back booster system we could have greatly expanded commercial space tourism because of the reduced cost of flying to space. This aspect alone could have greatly expanded space activity.

The success of the X Prize to stimulate commercial tourism flights to sub-orbital space has proven the desire of people to go to space. This will eventually lead to routine orbital space tourism. But it would have happened fifteen to twenty years ago if we had developed the fly-back booster. Space hotels would now be in orbit.

With low-cost routine flight to space we would have been able to afford more programs like the Hubble Telescope. We could have launched the hardware to make return trips to the moon or set up lunar habitats. With routine space launches we could have launched the hardware needed to build demonstration systems for solar power satellites. Modifications could have been made to convert the orbiters to fully reusable vehicles. New stretched S-IC's built with tooling that already existed, could have been built with up rated F-1 engines to increase payload and decrease operational costs. This could have happened just as Boeing took a small twin engine 737 concept with limited range and passenger capacity to become the most versatile biggest selling passenger airplane in history. Commercial versions of the Space Shuttle would be flying paying passengers to space in large numbers. It would have sparked the development of even lower operational cost systems and space would be a destination goal with achievable cost for many people of the world, instead of the very few and very rich. Our world would have been so different we wouldn't even recognize it today. It could have been achieved with a NASA budget not much different than what was actually spent. The money would have gone to operating the system and not for throw-away hardware and supporting a standing army of personnel that stand around waiting for the system to fly.

Probably the most damning indictment of NASA that has evolved over the Shuttle years is the loss of confidence of the people of this country and the world that we could build a low-cost dependable launch system. People hold their breath each time the Shuttle flies wondering if there will be another accident. It is becoming clear that NASA changed over the years from a dynamic, talented, technology smart organization, to a typical government bureaucracy that cannot seem to make tough decisions. The NASA management is much more interested in maintaining the status quo of the various centers than in accomplishing solid goals.

NASA expanded their image of ineptness with their abortive attempt to build the X-33 single stage to orbit reusable system. It was doomed from the beginning. To make a single stage to orbit system work requires very advanced light weight material that stretches the technology limits that exist today. Even if it could be built the payload fraction of a rocket powered single stage to orbit is very small and requires a very large vehicle that needs to be fueled with hydrogen. As a result the cost of fuel alone is enough to make the cost per flight high. The launch vehicle studies conducted over the years all conclude that with current technology, two stage vehicles are the lowest cost solution for low earth orbit and that the higher density of hydrocarbon fuel makes it the best engineering solution for the first stage. The upper stage benefits greatly from using hydrogen. Unfortunately, in recent years NASA has been much more intrigued with high technology gimmicks than they have been in practical solutions. The leadership that existed in the Saturn/Apollo era is no longer there.

Change may be on the way however. The latest Director of NASA appointed by Bush was removed from office shortly after the Obama administration took over. A new wind may be blowing at NASA in the future.

• • • • • • • • • • • •

In 1978 and 1979, we were able to convince the House of Representatives to pass a Solar Power Satellite Development Bill. It did not reach the floor of the Senate in 1978 but in 1979 hearings were held in both the House and Senate. The problem arose in the Senate hearings when the Department of Energy testified that if the bill was passed that President Carter would veto it. That was enough to stop a vote on the bill. If it had been passed by the Senate and signed by the President it would have provided the funds to continue development of the technology needed to carry the program forward. It would not have been a commitment for full scale development of the system, but it would have provided the funds and authority to proceed to a level of definition that would have made full scale development the next logical step.

If this had happened the commitment to full scale development would probably have occurred in 1983 or 84. Full scale development would have taken about 8 years to deploy the first satellite so electricity would have been flowing from space by 1992. Whether this would have been a government initiative or a commercial venture is an open question. However, if the government had funded expanded technology studies in 1980 there would have been a sufficiently sound basis for the commercial world to finance the development. This would have been particularly true if the government was willing to provide loan guarantees and purchase the first satellite for a government owned utility such as Bonneville Power. With commercial power flowing into the electrical grid additional orders would have emerged, particularly from other nations such as China, Japan or India. The early orders would have been made to fill a growing demand for electricity. As time went on and with the growing realization that global warming was a real danger, satellites would have been ordered to replace coal burning power plants.

What if we had tapped into an energy source with essentially unlimited capacity 24 hours a day? We would no longer need to worry about our energy life blood being oil from the Middle East. We would have started the conversion from oil to electricity from space to provide the energy to power our high technology world. It could have meant that our presence in the Middle East was reduced to the point that the terrorist attack of September 11, 2001 would not have occurred and the War in Afghanistan and Iraq would not have been justified.

What if there had been an explosion of high paying jobs in an industry that would support American workers? We could be exporting electricity in huge quantities that could reverse our massive foreign trade deficit. Instead of importing foreign oil we would be exporting solar energy from space and healing our environment.

What if we could now be involved in the exciting task of changing our energy paradigm from oil to electricity from space? It would be the beginning of the era of solar energy from space. What if we had gone ahead?

What If?

7   The Solution

Every President since Nixon has talked about energy independence. The only problem is none of them had a solution for achieving true energy independence. It was a nice sounding goal but when it came to defining how they thought it could be achieved they offered generalities and left the details up to others. They also failed to ask for, or provided funds to develop an energy source that could make us energy independent.

Jimmy Carter had the best opportunity because of the crisis brought on by the OPEC oil embargo of 1973-74. It provided a real incentive and he recognized the situation and did focus an effort under the Department of Energy. However, his bias against large programs doomed the effort from the start. Because of his nuclear submarine background he was influenced by the nuclear advocates so the effort was further distorted. The Carter administration focused their efforts on distributed energy systems that by themselves, a step in the right direction, could not supply the huge amount of energy needed to power an industrialized nation. During his administration the country lost their sense of direction and optimism for the future. Part of this disillusionment was due to his lack of ability or vision to inspire the people to look forward to a better world. He lost the opportunity to change the world and was voted out of office.

Ronald Reagan became President and with his buoyant upbeat personality changed the atmosphere of the country during the years he was in office. However, his attention was directed towards our military capability and the Soviet Union. Reagan's approach to energy was to let market forces do the work, not the government. By the time he entered office there wasn't much emphasis on new energy sources and he let big oil run the show. He even went so far as to have the solar panels, which had been installed by Carter's initiative, removed from the White House roof. Most of the alternative energy efforts by the government were left to drift with no real guidance or support.

The first George Bush had other problems to worry about and was more concerned with protecting Middle East oil than developing alternative energy sources. He was faced with the invasion of Kuwait by Iraq in their takeover of the Kuwait oil fields. He put together a coalition of countries whose armies drove the Iraqis out of Kuwait. During the years of his presidency we were sailing in the South Pacific. A lot of the time we were on remote islands that had very little contact with the outside world. We were not able to follow the details of what happened during that period, except we found that little progress had been made in solving the energy problem when we returned to the United States in 1993.

When Bill Clinton and Al Gore took office in 1993 I thought we might have a chance to bring solar power satellites to the forefront because of Gore's interest in the environment. I had returned from cruising to write **SUNPOWER: The Global Solution For the Coming Energy Crisis** and to form Solar Space Industries to promote solar power satellites. However, I was unable to establish contact with their administration. In the process of forming Solar Space Industries, I re-established contact with many of my former collogues and fellow advocates of space power. This included Dr. Peter Glaser the inventor of the concept. He told me he had briefed Al Gore when he was in the Senate. Unfortunately, Gore did not support the concept at that time. It is possible that he may have changed his mind since he has written his book, **An Inconvenient Truth**, which makes a profound case for global warming and its consequences. Solar power satellites would go far in eliminating the production of greenhouse gases.

George W. Bush is an oil man. He believes energy independence is drilling for oil in the Alaska wilderness and offshore. He is the latest in a line of administrations that have not faced the realities of the situation. George Bush's denial of global warming and its cause will one day be seen as one of the darkest periods in American and world history. He will probably be recognized in the future as the worst American President in history. He started a senseless war in Iraq, costing a great many lives, and hundreds of billions of dollars. He has lost respect for America throughout the world, brought on economic chaos, failed to recognize the damage to the world of global warming, and failed to promote the development of alternative energy systems. His administration squandered a budget surplus and created the greatest deficit in the nation's history.

We've had enough of politicians who failed to recognize the seriousness of our problems, just skirting around the fringes instead of aggressively seeking solutions. The issues are clearly evident. We need to: (1) find an energy system to replace oil; (2) stop global warming; (3) get out of the Middle East and let them determine their own future. And now, starting in the fall of 2008 we have another problem facing the world, a massive economic recession.

We also elected a new President. President Barack Obama has established a dynamic new administration that is aggressively attacking the problems that we face. He has the vision and leadership ability to bring us energy independence. The problems he faces are enormous, but there is also an enormous opportunity to change the paradigm of the world from an oil based economy to an electricity based economy. People have been searching for a Silver Bullet that would solve these problems and bring massive economic growth to the world.

There is a Silver Bullet. Developing solar power satellites is a clear, audacious, solution. *It is a new plan for a new era*. With their capacity to ultimately supply the entire world with low-cost energy as far into the future as the sun shines and life on earth exists. They would eliminate the need to use Middle East oil. They do not produce any atmospheric emissions to pollute our world. It is not a quick solution, but it can have immediate benefits that ultimately expand far into the future. It will take a huge initial investment by government and industry and many years to build all of the power plants required, but, the long-term benefits are enormous with wealth flowing from the sky, and may include the survival of mankind as we know it.

We have had leaders in the past who had the vision and courage to build for the future. Our founding fathers built a free nation; Lincoln freed the slaves and kept the nation whole. Roosevelt brought us out of the great depression, prepared us for World War II and led us to victory. Eisenhower created the interstate highway system. Kennedy sent men to the moon. Reagan brought down the Soviet Union and ended the cold war. Clinton created a budget surplus. Now Obama has the opportunity to lead us into a new energy era of prosperity and peace. Government does not have to foot the entire bill, but it does have to provide the leadership and co-ordination with other nations of the world to make the development possible. Let us not fumble the chance.

• • • • • • • • • • • •

"Why," you might ask, "are solar power satellites the right solution to replace oil?" Let me start by going back to the criteria for a new energy system. They are:

1. *Low-cost*
2. *Nondepletable*
3. *Environmentally clean*
4. *Available to everyone*
5. *In a usable form*

Solar power satellites must meet all of these criteria if they are in fact a true solution for the fourth energy era. Let's start with, *Low-cost*. Probably the main reason that solar power satellites have not yet been developed is because of the high initial cost of developing the system and the required space infrastructure needed for their launch and assembly. However, this is not the cost that counts in the long term. The cost of the electric energy that is delivered to the users is what's important. The high initial cost is associated with the fact that solar power satellites are only commercially viable when built in large scale. But this also means that because of the large scale they have the potential of producing very large quantities of electric energy and therefore producing a huge revenue stream. The development costs are mainly associated with developing the reusable launch system and the supporting space infrastructure. The satellite itself uses technology that has been in use on the earth for many years. The development costs are mainly associated with integrating them into a new system. Development costs, though high, can readily be amortized along with the cost of the satellites by the sale of electricity. This will occur in the same way that passengers buying tickets to fly on commercial airlines provide the funds to pay for purchasing the airplane and the costs of operating it. The cost of the airplane includes the cost of the airplane and the cost of developing it spread over many airplanes.

The cost of the electricity produced by the satellite will include the capital cost of building the satellite amortized over a period of years, the cost of maintenance and operation of the satellite and its ground receiver, and profit. Because the satellites can be built to last indefinitely, the cost of electricity drops to the cost of maintenance, operations, and profit after the capital cost is paid off. There is no cost for fuel as there is with coal, oil, natural gas, or nuclear power plants. Solar power satellites can be compared to hydroelectric dams in that there is no cost for fuel. To show you the potential cost picture of a 5 gigawatt satellite let us compare it to Grand Coulee Dam. It produces about 5 gigawatts of electricity over a year. Its cost to build was nearly the same as the cost of a 5 gigawatt satellite when you bring the cost into today's dollars. Today's cost of generating electricity by Grand Coulee Dam is among the lowest in the world.

As I look back to those days watching the dam being built during a period when jobs were so important to so many and realize what building power plants like that in space today could do for us, I ache to get on with the task. It was a massive task just as building giant satellites will be a massive task, but the benefits will also be massive as they were and still are from Grand Coulee Dam.

The long range cost of electricity from solar power satellites can be as low as hydroelectric dams and there is no threat of inflation due to increased cost of fuel. There is no fuel cost. In addition the output of a satellite is full power 24 hours a day for over 99% of the time. Power interruptions due to passing through the shadow of the earth will occur during the equinox periods of spring and fall at local midnight. It starts with a minute or two twenty days before the equinox, then increases each night until the equinox when it is a maximum of 72 minutes. It follows the same pattern in reverse over the next twenty days. Fortunately, the lowest energy demand tends to occur in the middle of the night during the equinox periods. In addition, if we are relying on satellite electricity as our primary source of energy, there will be many satellites spaced across the nation in geosynchronous orbit. They would not all be in the shadow at the same time so with grid inter-ties we would not even know when they were off line.

I will discuss more about cost in later chapters, but the point here is that energy cost can be very low in the long term, if we are willing to make the initial investment.

Sunlight is **Nondepletable**. So this is one of the major benefits of solar power satellites. They tap into the sunlight that normally streams right past the earth. It is available as long as the sun shines and when it stops we will already be gone. For us it is the ultimate energy source. Many satellites can be placed in the 165,000 mile long circumference of geosynchronous orbit. A day may come in the far distant future when we will have built a continuous ring at geosynchronous orbit circling the earth like the rings around Saturn. And in the near future the functions now being performed by communication satellites, observation satellites, and others can be integrated on the solar power satellites at much lower costs.

The most pressing criterion that we need to accomplish as we watch our world plunge into global warming, is it must be **Environmentally Clean**. Solar power satellites will not produce any atmospheric emissions as a by-product of generating energy. Since the energy is generated in space only the useful energy is transmitted to the earth in the form of a wireless power transmission energy beam. The radio frequency beam is converted back to electricity on the ground receiver with about 90% efficiency, so a minimal amount of waste energy is heat.

The launch of the satellite hardware will likely use a hydrocarbon fuel for the first stage and hydrogen for the second and will produce some carbon dioxide from burning the hydrocarbon.

The amount will be insignificant compared to the amount created by the fuels being burned to generate power and fuel our transportation system. The most desirable hydrocarbon first stage fuel would be liquefied methane. When the second stage burns hydrogen it produces water vapor as its exhaust. However, today hydrogen is extracted from hydrocarbon fuels which does free carbon dioxide gas into the air. After the early satellites are operational, low-cost electricity can be used to produce hydrogen from water and thus eliminate this source of carbon dioxide. Eventually the first stage launch vehicles can be converted to hydrogen fuel as well.

Another concern about the environmental effects of building solar power satellites is the energy required to manufacture the solar cells and to launch them. To give you a reference point, today it takes over two years to pay back the energy required to make a silicon solar cell used in a terrestrial solar panel. By comparison it would take only 6 months to pay back the energy used to manufacture and launch a silicon solar cell panel that is used on a solar power satellite. The difference is due to the greatly increased sunlight at geosynchronous orbit. This energy pay-back time will be reduced even further for thin film cells which are the likely candidates for the satellites today.

Many people wonder about the safety and environmental impact of the energy beam coming from the satellites. In their new look studies from 1995 to 2002 NASA looked at many alternative orbits, configurations, and energy transfer methods. This included the use of lasers to transmit energy to the earth. However, the studies of the 1970's concentrated on phased array radio frequency transmitters. They can operate in all weather conditions at an overall efficiency of 60 to 70% of electricity on the satellite to electricity into the grid on the earth. With today's technology that can be improved. In the definition studies we restricted the maximum energy density in the beam to 230 watts per square meter. This is less than one quarter of the energy density of bright sunlight on earth. Exhaustive studies on the effects of radio frequency energy on all life forms has subsequently found that the threshold of damage to living cells, which is due solely to heating, starts to occur at 1000 watts per square meter. As a result the energy density in the beam can be increased to 500 watts per square meter or maybe even 750 watts per square meter and still be safe for all life forms.

At these levels of energy density it is safe for people if an airplane flew through the beam or birds wandered into it or bees came looking for honey. The questions that people ask me when I talk about the energy beam usually focus around their worry of being cooked. I can assure them without any doubt that it is not something that they need to worry about.

The receiving antenna which is a rectifying antenna would be raised above the ground allowing 80% of the sunlight to pass through, but stops over 99% of the radio frequency energy, so that the land under the rectenna can be used for cattle grazing, or farming. One elegant design would be for the rectenna elements to be imbedded in the glass roofs of green houses. Another optional use is to mount terrestrial solar cell panels under the rectenna using the pass-through sunlight to supply daytime peak load electricity. Imagine – electricity from solar satellites and terrestrial solar panels at the same time. How about that for efficiency of effort? One other possibility is to use the land under the rectenna to grow algae for bio-fuel. This would allow the satellite system to also solve the problem of powering airplanes, ships and trucks.

The use of laser beam power transmission has also received considerable attention. I attended the State of Space Solar Power Technology conference sponsored by the Air Force Research Laboratory in October, 2008 in Orlando Florida. This conference emphasized the need of the military for energy availability at remote and advanced sites. The use of space based energy systems that could beam energy to these sites could become very important to them. The magnitude of the energy needed is much less than a full sized satellite that would be appropriate for commercial use and they also need a much smaller receiver area for the energy coming to earth. The use of lasers would fill this need. The work that has been done covered both the formation of the laser beam and the receivers which would be composed of high efficiency solar cells that matched the light frequency of the beam. One of the benefits of lasers is the ability to use small solar arrays in space to drive one or more lasers that would beam their energy to the earth. These could be clustered with other arrays and lasers to bring the energy level to the desired magnitude. The spot size on the ground could be kept to a very small area. The lasers would be subject to interruption by cloud cover, but the other benefits for the military use or potentially small commercial use make this a good solution for them.

The problems associated with laser beam transmission through the atmosphere and clouds in addition to the perceived safety issues of lasers will likely limit them to specialty uses, but not be the primary energy transfer method for large scale commercial use. The wireless radio frequency system will likely remain the baseline approach for large scale satellites. The baseline radio frequency used in the System Definition Studies was 2.45 gigahertz, the same as a microwave oven, but there are other options.

The bottom line is that the energy beam will be safe for all life forms and there are good solutions for the use of the land under the rectenna. Even without dual use of the land, the total land use is less than other existing energy systems such as coal, nuclear, and hydroelectric.

Solar power satellites will be environmentally clean.

By placing the satellites in space around the equator energy can be *Available to Everyone*. Today our power plants are located in the part of the country where the energy is used. We do have distribution grids that span long distances, but not oceans. Each country has its own power plants. In the case of many developing countries they cannot afford the capital cost of power plants. The fuel used tends to be the fuel that is available in each country or that they can import. Solar power satellites are unique in that they can be positioned to deliver electricity to any nation on earth and they do not have to be owned by the country that receives the power. All they would need is a rectenna, which will be about one third or less the cost of a satellite. They then can buy the electricity from the satellite. Whoever owns the satellite will be an energy exporter. The net result is electricity can be made available to everyone.

The energy output of solar power satellites is electricity. It is the most usable form of energy we have. It can be used to do nearly any energy task we ask of it. It can power our high technology world. It can provide light, heat, motors, communication systems, and many forms of transportation. It can provide the energy to purify water and make hydrogen out of water without environmental pollution. If we stop using oil and natural gas for most transportation systems the remaining reserves could supply the needs of airplanes and ships until new technologies are developed to take their place.

Solar power satellites produce energy *In a Usable Form*.

As you can see solar power satellites meet all of the criteria for the energy system for the future. Now let us explore how they can solve the problems we face today.

• • • • • • • • • • • •

With world oil production approaching or at its peak we will experience increasing shortages and the inevitable result of increasing costs of gasoline, diesel, and airplane fuel. We experienced dramatic increases when supply was interrupted by hurricanes Katrina, Rita, and Wilma. That was just a foretaste of what is happening as world demand continues to rise and supplies can no longer keep up with demand. Don't be misled by temporary drops in the price as supplies seem to become sufficient to meet demand. When we are near peak world production there will be periods of sufficient supply and then shortages that will cause price increases. The higher price will cause a depressing pressure on use that can then create temporary surpluses. We have seen this happen in the summer of 2008 as oil prices soared to $148 dollars a barrel and then dropped back to the $40 to $50 dollars a barrel range. The cycle of shortages and surpluses will become shorter and each oscillation will result in ever increasing costs. We will only be able

to verify that we have passed the actual peak of oil production at least a year or two after it actually occurs. In the meantime the price will continue to increase and that will force us to conserve. It has only been a short time that oil prices have surged, but the cost has already caused serious impact on our economy. Energy costs are one of the fundamental drivers of inflation and the cost of living, as they impact the cost of everything we buy or do.

New exploration for oil will undoubtedly find some new supplies, but in the meantime older fields are in serious decline. The new discoveries, even if they are huge, can only delay the peak, not prevent it.

We have already waited much too long to develop a replacement for oil. It takes about forty years to bring a new energy system to the point it can supply massive amounts of energy. All of the alternative energy systems we have developed or are developing have limited capacity. Only solar power satellites have the massive capacity potential to really replace oil. So solar power satellites can solve our problem of replacing oil, but we need to act now!

• • • • • • • • • • • •

Global warming is a phenomenon that was denied by many people and many of our leaders for years. Even many scientists didn't believe that we could cause global warming. However, the evidence of it happening and its cause is becoming overwhelming as we continue our journey into the 21st century. Green house gases led by carbon dioxide are the culprits. There has been much talk about reducing green house gases. The Kyoto protocol, that the United States has refused to sign, was developed to attack the problem, but all the efforts to date have been minuscule compared to the magnitude of carbon dioxide we are dumping into the atmosphere. This is a problem that may be compounded by deforestation in the Amazon Basin. We have to stop burning vast amounts of fossil fuels if we hope to stop global warming. The connection between fossil fuel energy and global warming is now clear.

The one clear solution to both the problem of oil and global warming is solar power satellites. They have the potential capacity to replace oil as the primary energy source and to replace coal and natural gas for generating electricity without producing carbon dioxide as a by-product. It cannot be done overnight but it can be done. Solar energy from space can provide the electricity to charge batteries for automobiles and to produce hydrogen from water without atmospheric pollution so that we can move toward a battery and hydrogen powered transportation system. Because the United States uses 25% of the world's energy we are in a position to have the biggest impact on reducing carbon dioxide emissions. But solar power satellites can supply other countries with energy as well so their ability to stop carbon dioxide emissions will be worldwide.

I expect by now you are thinking; "How in the world can we afford to build all these enormous new power plants in space?" My response is; "How can we afford not to?" A more reasoned answer is that they will be commercially viable power plants that will pay for themselves over a period of years. The big issue will be the cost of shutting down existing polluting coal plants and scrapping our gas guzzling automobiles. Internal combustion engines can run on hydrogen if properly modified so many of our cars will probably be around for a while. The railroad system will need to be expanded and converted to electric locomotives. The expansion will be needed to take over much of the current trucking business. It is obvious that we are in for a major paradigm shift that will cause many businesses to disappear, but at the same time there will be tremendous opportunity for new businesses to evolve. The task of building, launching, and assembling the satellite hardware will create entire new industries and jobs. The pace of the shift will be controlled to a great extent by our ability to fund the changes, but if we don't, the costs of not doing it will be catastrophic for our children and their children. Are you willing to take that failure to your grave?

There is a solution. Let us not miss the opportunity of implementing it.

• • • • • • • • • • • •

We are embroiled in wars in the Middle East that have cost many lives.

Theirs is a culture and a religion that is different from ours that we do not fully understand. We are there because we are so dependent on oil. Other reasons have been used by our government to justify our interference, but the bottom line is we need their oil. Our presence triggered the 9/11 attack. Imaginary weapons of mass destruction justified our invasion of Iraq and we are trying to impose a democratic form of government on them that they do not want. We have lost the respect of the world with our high-handed egotistical stance.

We need to free ourselves from dependence on Middle East oil and get out of the region and let them determine their own future. The key to making that practical is to replace oil as our primary energy source.

So again we come to the one solution that can solve all three problems. If we develop solar power satellites, we can replace oil with energy from the sun, stop global warming, and eliminate our dependence on Middle East oil and revitalize our economy. As my grandson would say; "That's a no brainer." If it was only that easy for people to see that there is a solution and demand that it be implemented.

8   The Fourth Energy Era

I have talked about the different eras of energy we have experienced, first wood, then coal, and now the era of oil. The era of oil will not be over until we have moved on to a new primary energy source. As I have postulated this energy source will be solar power satellites, if we have the courage and foresight to develop them. So far there has not been a single kilowatt of electric energy transmitted from space to the earth to power anything. However, that doesn't mean that radio frequency energy isn't flowing from geosynchronous orbit to the earth for our use here. It just isn't energy that powers machines or lights. It is signal strength energy that transmits to our television sets, telephones, weather stations, and earth observations. The basic concept is similar to what will happen when we transmit energy, except it will be wireless power transmission of much greater magnitude.

When the next energy era will start is uncertain, but it needs to start soon if we are to solve our problems and save our planet. In the next chapter I will discuss a potential plan for how we can go about their development. But first let's look at what the satellites will probably look like and how they will work.

By far the most comprehensive studies of solar power satellites were conducted by the System Definition Studies sponsored by the Department of Energy and NASA in the 1977 to 1980 time period. Many people assume that since they were so long ago they are no longer valid and that technology has passed them by. However, engineering principles don't change, they are timeless. But there have been technology changes that resulted in improvements that can be made in all of the systems, many of which I will cover here and in later chapters, but the fundamentals are the same. As a result I will start with the basic configuration of the satellite system as it was developed in the System Definition Studies. The new look studies sponsored by NASA in the 1995-2002 time period had a different focus. They were looking for a way to develop the system without the need for a massive initial investment. They looked at alternate orbits, smaller satellites, laser energy beaming, alternative configurations and launch approaches. They followed the typical NASA approach after the Saturn/Apollo era of being more focused on high technology than commercially viable concepts. They did not include the definition of a new reusable launch system which is essential to the development of solar power satellites and is the first system that must be developed. Other configurations have been proposed recently but they have not had the engineering definition to make them viable at this time. As a result I am going to stick to the basic configuration used in the reference system which was thoroughly engineered. When the time comes to do the actual design the engineers will evaluate all approaches and select the best that is available at that time to actually build.

Figure 3: Solar Power Satellite Configuration
from the Definition Studies

The concept of solar power satellites is based on three basic functions. The first is to intercept sunlight in orbit around the earth and convert it into electricity. The second is to convert the electricity into an energy form that can be transmitted to the earth. The third function is to receive the energy on the earth and convert it back into electricity. The ideal orbit for the satellite is geosynchronous orbit. This is the orbit over the equator at an altitude of 22,300 miles (36,000 kilometers). A satellite in this orbit has a period of rotation around the earth that is exactly equal to the 24 hours rotational period of the earth. As a result the satellites stay over one location at all times.

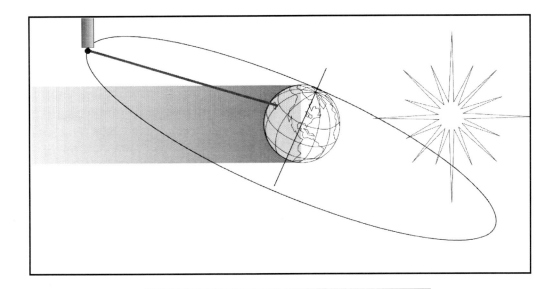

Figure 4: The Satellite is in the sunlight
over 99% of the year

Because the earth's axis is inclined 23° from the plane of its orbit around the sun, the satellite passes either above or below the shadow of the earth for most of the year. As I mentioned earlier it only passes through the shadow when the earth passes through that portion of its orbit when its axis is perpendicular to the sun during the equinox periods. At these times it is tilted forward and backward in relationship to the earth's orbit around the sun. It is in shadow less than 1% of the year. Many of our communication satellites and television broadcast satellites are currently in geosynchronous orbit.

• • • • • • • • • • • •

There are many variations on how to convert sunlight into electricity. One of the earliest concepts developed by Boeing in the mid 1970's was to use a series of mirrors arranged in a dish form to concentrate sunlight into a cavity absorber. This heated a gas to a high temperature that then ran a turbine that powered a generator. The conversion efficiency was quite high and the concept was called PowerSat. During the DOE/NASA studies we decided that even though the efficiency was high, the long term reliability and life would not be as good as solar cells. Solar cells convert sunlight to electricity directly without any moving parts. There are many different materials that can be used to make solar cells. Their costs, weights, and efficiencies vary significantly. The decision on which is best is dependent on a trade-off of these factors along with their life expectancy and ease of installation. In addition, cells can be used in flat planer arrays, like most terrestrial solar cell panels you see mounted on roof tops, or fewer cells can be arranged with mirrors that concentrate the sunlight to significantly increase the output.

The Rockwell International configuration used a concentration system with their solar cells. Boeing's final configuration for the 5 gigawatt reference design used a planer array of thin, single crystal silicon cells with an efficiency of 16.5%. This system was selected as the basic reference design. This is a huge satellite with nearly twenty square miles of solar cells. As the efficiency of the solar cells become higher the size of the satellite will be reduced for the same output. Also as the weight of the cells is reduced the weight of the satellite is reduced and the launch cost is lowered. The satellites can also be made with smaller output. There are several considerations to take into account when making the decision on what size satellites will be the best solution. This decision needs to take into account the wireless power transmission beam, the adaptability of the earth distribution grid, the energy need on the earth, and the resulting energy cost on the earth. As long as radio frequency energy is used for the wireless power transmission beam economics will favor the larger satellites. This is due to the physics of the energy beam which we will look at in greater depth later.

The physical shape of the reference design was a rectangle with the solar cells arranged in square bays. The satellite was positioned so that the cells always faced the sun. This meant that the transmitting antenna needed to rotate one revolution each day to keep it pointed towards the earth. To accomplish this it was located at one end of the rectangular satellite and looks like a little head on a giant body. The rectangular shape was selected for the ease of space assembly using robotic techniques. The NASA configuration from the new look studies was called the Power Tower and was composed of circular solar cell arrays supported off of a central structural column. I assume that the circular arrays were designed for deployment as a unit in space. There had to be some logical reason to complicate the design by using a circle which does not lend itself to repetitive mass production of solar cells or robotic assembly. I would expect that when commercial satellites are being built they will be in some form of squares or rectangles.

• • • • • • • • • • • •

The heart of the system is the power conversion from sunlight to electricity on the satellite, but what makes it work for us is the transmission system that brings the energy to the earth. In the reference design, the wireless power transmission system was based on a phased array transmitter that is pointed towards the earth receiver, but the final guidance of the beam is done electronically. The transmitter is made up of sub-arrays composed of slotted wave guides. Each sub-array has the phase of its frequency controlled to form a cohesive beam that is steered by also controlling the phasing. The design is such that the fine steering is controlled from the ground so that the beam cannot wander from the receiver. The electrical energy was converted to radio frequency by high power klystrons similar to those used by radar systems. These are very efficient, but need supplemental cooling systems due to their high power. The frequency selected is 2.45 gigahertz (2,450 megahertz) which is the frequency used in microwave ovens. After the development of the reference system design, Bill Brown, who invented the concept while with Raytheon, came up with a simple method of converting microwave oven magnetrons from oscillators to phased lock amplifiers. This raised their efficiency to the level that because of their low-cost and smaller size could make a significant improvement in the system. In addition the Japanese performed an in-space transmission test of energy using solid state converters. Technology continues to advance so there will undoubtedly be other options that could be used if they can improve on the concept. But the point is that existing technology can do the job quite well.

As I mentioned before the wireless transmission beam influences the most economical size for the satellite. This is because the diameter of the beam when it reaches the earth is determined by the frequency in the beam, the distance the beam travels to reach the earth, and the diameter of the transmitter. It is independent of the magnitude of energy transmitted. These factors are basic to the principles of physics of the beam. When the frequency is higher the diameter on the ground is smaller. Also in order to make the diameter smaller on the ground with the same frequency the transmitting antenna must be increased in diameter. The distance from geosynchronous orbit is fixed so that factor is constant. The starting point is determining the energy density in the beam when it reaches the earth. In our System Definition Studies that was set at a maximum of 230 watts per square meter. That was originally set because of a theory developed at that time that it was the maximum that could be transmitted through the atmosphere without triggering run-away heating in the ionosphere. However, testing that was done later from Arecibo and Platteville earth based radio transmitters determined that this was not the case. In any event we used this limit. Using 2.45 gigahertz as the baseline frequency and a transmitter diameter of 1 kilometer resulted in a power level of 5 gigawatts as the satellite's ground output. This was the size selected as the baseline for the studies.

The 2.45 gigahertz frequency passes through the atmosphere with very little loss in power and even in severe rain storms would only lose about 10% of its energy. This frequency is in the Industrial Scientific & Medical (ISM) band and is used throughout the world in microwave ovens. If the other candidate frequency of 5.8 gigahertz is used the atmospheric losses increase. Frequency selection and dedication to energy transmission will involve international negotiations and agreements. The communication world will oppose any selection, but as Dr. Peter Glaser once said; "When push comes to shove, energy will win." We need clean energy to survive and we can work around communication problems which all of the studies found would be minimal.

The trades that need to be done before determining the satellite size must consider the relative sizes and cost of the transmitter and the rectenna. The evidence appears to be conclusive that 500 watts per square meter is a safe level. It may be that even higher levels could be safe for brief exposure times that would only occur on rare occasions, such as a small airplane inadvertently flying through the beam. It may be possible to change the energy contour in the beam from a Gaussian distribution which is bell shaped (see illustration in chapter 13) to more of a barrel shaped that would increase the total energy transmitted without increasing the maximum energy density or size on the ground.

In addition to the technologies I have talked about there are more options available today such as solid state devises for converting the electricity to radio frequencies. NASA studied laser beams for the power transmission during the New Look Studies. This effort has been continued and is of particular interest to the military, as I discussed in Chapter 7. Laser power transmission has the advantage of being able to transmit small amounts of power to the earth to small receivers. The disadvantages are that clouds can block the transmission and in the case of high energy levels there will be a question of safety. The use of radio frequency wireless power transmission is a clear winner for large commercial energy satellites at this time.

• • • • • • • • • • • • •

The third element of the system is the receiving antenna on the ground. This is the element that brings the energy from the satellite to us in a useful form. We have Bill Brown to thank for this highly efficient system. It is actually very simple in concept. The receiver would be composed of half-wave dipole antenna elements that are just short pieces of fine wire spaced a few inches apart. What makes it able to deliver electricity are diodes that rectify (change) the received radio frequency energy into DC electricity. In the original design there was one diode for each half-wave dipole. However, with later development one diode can rectify the energy from several antenna elements. They can be interconnected in series/parallel streams to gather the energy at appropriate voltages to feed into inverters that change the electricity to AC current. The inverters would be the same concept that is used in terrestrial solar arrays to convert the DC electricity into AC current that can be fed into the existing electrical grid. The efficiency of this process can exceed 90%. With these technologies the overall efficiency of the power transmission from electricity on the satellite to electricity into the grid will be 60 to 70%.

One of the startling features of solar power satellites is they will have the highest efficiency of any energy generating system. They will have an overall efficiency of over 8% from sunlight in space to useful energy on earth. With new technology the efficiency could go much higher. There is no other energy source that even comes close to this level of overall efficiency. All of our stored energy sources such as coal, oil, and natural gas start with the 1% efficiency of photosynthesis and go downhill from there. They all have to go through a thermal cycle to convert the heat of burning into first mechanical energy and then to electricity. So the very best fossil fuel systems are only about 1/2 of 1% percent efficient overall. When we count overall conversion efficiency of nuclear fission it is about 1% efficient. Terrestrial solar can be in the neighborhood of 3% overall efficient. The lower efficiency of terrestrial systems compared to space systems is because of the day/night cycle and atmospheric losses. Their short term efficiency during the day can be considerably higher, but they stop functioning at night. With wind energy it is nearly impossible to establish an overall efficiency number. Nuclear fusion, if it is ever developed on the earth, will probably be less efficient than fission because of the extreme temperatures and pressures involved. It is important to understand that solar power satellites are the direct

conversion system for converting the fusion energy of the sun to useful energy on the earth. Apparently there is a natural system that has higher efficiency in converting sunlight into growing plants than photosynthesis. It is Purple Pond Scum which apparently is very efficient in converting solar energy. This and other algae are being studied by scientists to turn into fuels. The amount of money being invested by venture capitalists in research on algae based fuel is now skyrocketing.

One other potential improvement in satellite efficiency would be if the concept of the rectenna, which rectifies radio frequency to electricity, can be escalated from radio frequency to the frequency of light. Then the satellite efficiency would jump by half an order of magnitude. There is research going on to develop this concept. By using nano-technology the light frequency may be directly rectified to electricity at 70% to 90% efficiency. This would replace solar cells to convert sunlight into electricity and dramatically reduce the size of the satellites.

Because the satellites will be in geosynchronous orbit over the equator and most of the populated countries are north of the equator with some south of it, the energy beam will be coming to the earth at an angle. The angle of the beam from vertical when it reaches the earth is the same as the degrees of latitude of the receiving rectenna. As a result the beam's footprint will be an oval instead of a circle. The receiver will be composed of rows of antenna elements that will look like wire mesh fencing material. They will be supported above the ground so that the mesh is perpendicular to the beam coming down from over the equator. The rectenna will capture over 99% of the energy in the beam while at the same time allowing 80% of the sunlight to pass through. An early test of the rectenna was conducted by NASA in 1975 at the Goldstone facility in California when they transmitted 30 kilowatts of energy over about a mile distance to a receiver. In that test the rectenna was about 72% efficient.

Because the energy density in the beam must be limited to safe levels the size of the rectenna is very large. If the maximum energy density is 230 watts per square meter the diameter at the earth is about 5 miles for a 5 gigawatt satellite. If the energy density is increased to 500 watts per square meter this would reduce the beam diameter to about 3 miles. This is a large area, but when compared to the land destroyed by strip mining coal for the same amount of energy over 40 years it is small in comparison. In addition the land under the rectenna can be used for other purposes. The challenge is to locate the receiver sites reasonably close to the large users which means cities. This is not a hard criterion, but would help minimize long transmission lines. Using Grand Coulee Dam as an example again, it delivers power into the Bonneville Power grid that gets distributed throughout the Northwest and is also inter-tied to California. So we can transmit electrical energy over long distances, but it would be better if we did not have to do that. The expanded and modernized national electrical grid system being proposed by the Obama administration will be a big help.

• • • • • • • • • • • • •

Now you have an idea of how solar power satellites work. The next question is how will they usher in the fourth era of energy? The era of wood evolved because it was the natural fuel available nearly everywhere. People only needed to pick it up or cut it to convenient size for burning. The era of coal was started because wood became scarce in England and they were forced to find an alternate source of energy. The era of oil evolved because of its superior properties, abundance, and low-cost. The fourth era, solar energy from space will be triggered by the impending scarcity of oil and worldwide need to reduce the emission of carbon dioxide into the atmosphere to reverse global warming. The recent release of the **Stern Review: The Economics of Climate Change**, has brought to the attention of the world very serious global risks, and it demands an urgent global response. This independent review was commissioned by England's Chancellor of the Exchequer, reporting to both the Chancellor and the Prime Minister. It assesses the evidence and builds understanding of the economics of climate change due to global warming. It identifies the potential magnitude of the economic damage from climate change and the need to act quickly to limit the increase in greenhouse gases.

We now have the driving forces of the great economic impact of depleting oil reserves and global warming that will bring on the start of the fourth energy era. Both of these economic forces added to the financial collapse of 2008-2009 have the potential of sending the world into an economic dislocation that could eclipse what happened in the great depression of the twentieth century. Maybe the world needs that traumatic experience to force the necessary steps of changing our energy source. It is too late to avoid all of the negative effects, but we can minimize the impact if we act now. Solar power satellites can do for the world what Grand Coulee Dam did for the Pacific Northwest during the depression.

The logical sequence of events is for the initial satellites to be built to fill the expanding electrical energy demands of the United States and other growing economies and to slow the growth of greenhouse gases. This is particularly acute in the emerging economies of China and India. The immediate effect of developing the satellites will be the jobs that will be created to bring them about. By the time the early satellites start supplying electricity to the earth, world oil production will have passed its peak and the conversion from oil fueled systems to electricity will have begun. This will initiate a new wave of demand for more electricity. Along the way it is also important that capacity be added that can replace fossil fuel power plants, particularly coal so that their pollution of our atmosphere stops.

The pace of this transition will be controlled by two major factors. The first will be the ability of the industries of the world to manufacture and assemble the satellites. There are many facets of this process that will impact how quickly the process progresses. This includes the capacity of the space transportation system, the capacity of the production facilities for solar cells, antenna elements, satellite structure, and efficiency of the space assembly process. Massive new industries will be created to fill these needs that will require workers to fill the jobs. Much of the manufacturing functions will be automated to degrees beyond most industries today, but nevertheless the work force will be large. The secondary jobs this labor force will generate will add dramatically to the overall number of jobs that will be created.

The second major factor that will control how fast the satellites are built is the financial resources of the world. This will cause a paradigm shift of a magnitude that has not been seen before. It will be a transition period for the entire world as we change from oil to electricity from space. Since we will be starting so late the transition will have to take place as rapidly as possible to minimize the dramatic economic impact that the increasing shortage of oil and global warming will bring. The economic cost impact due to capital investment in the satellites will be large. Initially it will strain the resources of the world, but it will soon be off-set by the revenue of wealth streaming from the sky.

During this period oil will continue to be a major source of energy for transportation use for those who can still afford the cost. Other energy sources will also make significant contributions. There will not be any dramatic event that marks the transition from the era of oil, but one day the realization will come that oil no longer dominates the energy world and the fourth era, energy from space will be in effect.

9   A Development Plan

The great dilemma we face is how to start a solar power satellite program that can bring about a new energy era and solve our serious problems. The basic difficulty has been the high initial development cost. NASA tried to find a short-cut that would allow smaller systems so the initial costs could be reduced, but still development did not get started. The development costs include developing a new low-cost space transportation system and the entire space infrastructure needed to assemble and deploy the satellite, as well as the development of the satellite and wireless power transmission system. None of this existed when the original studies were conducted so the cost was very high.

The situation is much different now. In the ensuing years many things have changed. Other programs have sponsored research and development of several of the enabling technologies and some of the required infrastructure has been developed. Studies have continued in several countries outside of the United States and some limited activity is sustained by individuals and companies on their own funds within the United States.

It is essential that the United States be the leader of this endeavor. As history has so clearly shown, whatever nation develops and controls the world's primary energy source dominates the world economy. There is currently a great deal of interest in energy from space in India, China and Japan. Japan has even announced it plans to deploy a prototype in the not too distant future. If we lag in moving forward and one of the other nations take the lead we will suffer the consequences.

In 2000 I was asked to testify before the Subcommittee on Space and Aeronautics, United States House of Representatives Committee on Science, concerning solar power satellites. As part of my testimony I proposed an industry/government partnership plan for developing solar power satellites. Nothing has been done yet, but I still believe it is a good plan. Here I am going to include both my brief oral statement before the committee and the full text of my testimony from 2000 and let you see if you agree.

Summary of Statement of Ralph H. Nansen,

President Solar Space Industries

Before the House of Representatives Subcommittee on Space and Aeronautics

September 7, 2000

Mr. Chairman, Members of the Subcommittee, thank you for inviting me here today to testify about the feasibility of Space Solar Power. The last time I appeared before this Subcommittee was in 1978 when I accompanied the President of Boeing Aerospace Company to testify concerning the same subject. At that time I was Program Manager for Solar Power Satellites for the Boeing Company. Today I am retired from Boeing and am currently the President of Solar Space Industries, a company I formed in 1993 to promote the development of Solar Power Satellites. I also wrote an advocacy book about Solar Power Satellites in 1995 titled, **SUN POWER: The Global Solution for the Coming Energy Crisis***.*

Much has change in the last 22 years since I was last here, but one thing that hasn't changed is the fact that Solar Power Satellites are still not under development. However, the time is now right for their development to begin.

The studies conducted in the late 1970's determined technical feasibility and the potential promise of Solar Power Satellites for delivering, abundant, low-cost, nonpolluting electric energy to all the nations of the world. Studies since that time have reaffirmed this conclusion. In addition much of the infrastructure that did not exist in the 1970's has been developed for other programs, dramatically reducing the development costs.

A low-cost reusable space transportation system required for space solar power has not yet been developed. However, Solar Power Satellites would provide a large enough market to justify its development.

The need to develop space solar power is becoming more apparent as we see energy demand growing throughout the world, energy prices rapidly increasing, oil reserves dwindling, and the threat of global warming. Space solar power can solve these problems.

A potential interim step is transmitting energy from one location, that has excess energy capabilities, to another location on the earth by reflecting a wireless power transmission beam with a relay satellite in geosynchronous orbit. Because the relay satellite would be light in weight, it could be launched with existing expendable launch vehicles.

One of the key issues before us today is what should the government be doing about space solar power. The development of the system should primarily be a commercial development, however, because of the size of the program required and the international implications it should start as a government/industry partnership. The primary role of the government would be to provide leadership and seed money to initiate the program, coordinate international agreements, support the development of high technology multi-use infrastructure, establish tax and funding incentives, and assume the risk of buying the first operational satellite.

Industry can provide most of the developmental funding and be responsible for the design and development of the system. It is essential that the satellites and the space transportation system be developed in a commercial environment if they are to be viable commercial ventures.

The following steps by the government would bring this about.

1. Assign a lead agency within the government. The Department of Energy is the logical lead agency with NASA providing the primary technology support.

2. Fund a Ground Test Program to demonstrate the satellite functions of power generation, the wireless power transmission system, and integration of the energy into a utility grid. This program would also demonstrate the capability of relay satellite power transmission.

3. Obtain frequency allocation for wireless power transmission.

4. Pass commercial space tax incentive bills, like the Zero Gravity, Zero Tax bill.

5. Incorporate testing for solar power satellite technology into the plans for the International Space Station.

6. Continue technology development for reusable space transportation systems.

7. Consider the implementation of loan guarantees for commercial development of reusable space transportation systems.

8. Commit to the purchase of the first operational Solar Power Satellite.

With this plan implemented the commercial industry would have enough confidence to proceed with development. Most important of all is the fact that whatever nation develops and controls the next major energy source will dominate the economy of the world.

Full Testimony

The Technical Feasibility of Space Solar Power
Statement of Ralph H. Nansen,
President Solar Space Industries
Before the Subcommittee on Space and Aeronautics, United States House of Representatives Committee on Science
September 7, 2000

Abstract

The concept of solar energy generated in space for our use on the earth was first proposed by Dr. Peter Glaser in 1968 and has been studied extensively since then. The technology required for its development is known and solar power satellites have the potential of delivering abundant, low-cost, nonpolluting electricity to all the nations of the earth. Their development has not proceeded because of high initial development cost of a reusable heavy lift launch system and other supporting space infrastructure. The time is now right for the United States to lead the world in developing the system. Fossil fuel costs are rising as world demand is increasing while supplies are dwindling. In addition global warming highlights the need to reduce carbon emissions in the atmosphere. The development of solar power satellites can solve these problems and bring economic dominance to the nation that develops and owns the system. The government's role in this program should be to provide leadership, seed money, and incentives for commercial development. Specifically the funding of a small scale Ground Test Program over a three year period at a funding level of $30 million a year would demonstrate to the commercial community the viability of the system. This along with tax and other incentives will bring about commercial development and the resulting benefits to the United States and the other peoples of the world.

Introduction

The concept of solar power satellites was conceived by Dr. Peter Glaser in 1968, but it was made possible by the work of William Brown of Raytheon. The idea of generating electricity in space for use on the earth was treated as an unrealistic dream when it was first presented. However, a few individuals in NASA thought it ought to be investigated, so there were some low-level studies initiated to look at feasibility. They concluded the concept appeared to be technically feasible and the cost might be low enough to be competitive if the cost of space trans-

portation could be reduced significantly.

A study of Future Space Transportation Systems conducted by Boeing for NASA concluded that transportation costs could be lowered to very low levels with the right type of reusable launch vehicles. This opened the door to further studies of the system. When the OPEC oil embargo in 1973-74 triggered an energy crisis in the United States an effort to develop alternative energy sources, including solar power satellites, became a national priority.

In the late 1970's a broad-based Systems Definition Study was conducted under the joint auspices of DOE/NASA. The System Definition prime contractors were The Boeing Company and Rockwell International. I was the Program Manager for Boeing during this period. The studies which involved a large number of contractors and organizations concluded that solar power satellites were technically feasible and had the potential of being economically competitive. The problem was the huge cost of development and deployment of the system before producing significant revenues. There were also uncertainties on the level of technology maturity, infrastructure development, and cost estimates. As a result of these concerns, coupled with the political opposition from the nuclear industry, the government program was terminated in 1980.

Since 1980 organized activity to

study or develop solar power satellites has been limited. There was no US government sponsored work until NASA initiated their "New Look Studies" in the mid 1990's. Subsequently the Department of Energy abstained from any involvement. However, during this time the Japanese government and industry became interested in the concept. The Japanese updated the reference system design developed in the System Definition Studies in the late 1970's, conducted some limited testing and proposed a low orbit 10 megawatt demonstration satellite. Their effort has been curtailed by their economic problems. Interest by other nations has persisted, but only at low levels of activity. The overwhelming initial cost of development and deployment has remained the primary obstacle. Number one on the list of cost barriers is the cost of space transportation. Solar power satellites are only economically feasible if there is low-cost space transportation.

In spite of the lack of organized activity to develop solar power satellites much progress has been made. Most of the development that has occurred is in maturing technology of the subsystem elements and space infrastructure. This includes solar cells, power processors, wireless power transmission components, robotics, space habitation modules, reusable launch vehicle technology, and computational capability.

A companion program to solar

power satellites was investigated which would utilize the same concept of wireless power transmission to deliver electrical power from one location on the earth to another several thousand miles away. This could be accomplished by transmitting radio frequency energy in a wireless power transmission beam to a relay satellite in geosynchronous orbit, which would reflect the energy back to a receiver on the earth called a rectenna. This concept would allow transmission of excess energy at one location on the earth to another that needs the energy without the need to construct long distance power transmission lines. This is an attractive option. It does not require the development of new space transportation systems as the relay satellite, though large in diameter to reflect the wireless power transmission beam, can be light in weight. The only active systems required on the satellite are for pointing control and station keeping.

Over the last two decades knowledge of the potential of solar power satellites to provide our world with unlimited clean energy has drifted from the public conscience. Most people are unaware of the concept today.

This testimony is presented to provide a brief review of Solar Power Satellites, why they have not been developed , why the United States should develop them, what the situation is today, and of particular importance is the steps the United States Government can do now to speed their development,

What are Solar Power Satellites?

Solar power satellites as envisioned are large-scale power plants based in space in geosynchronous orbit. The satellites would be in the sunlight for over 99% of the year. They would only pass through the shadow of the earth for brief periods during the spring and fall equinoxes. Electric energy would be generated by vast arrays of solar cells converting sunlight to electricity. The electricity would be routed to a phased array transmitting antenna that would convert the electricity into radio frequency energy and transmit the energy in a wireless power transmission beam to an earth-based receiver. This receiver, called a rectenna, would convert the radio frequency back into DC electricity. Power processors would then convert the DC electricity to AC power for distribution on existing power grids.

The power output of each satellite studied during the System Definition Studies was 5 gigawatts (about equivalent to the output of Grand Coulee Dam). Smaller satellites are possible. However, smaller satellites still require large space transmitters which result in increased cost of the electric power delivered to the earth. One gigawatt output is probably the smallest practical size for geosynchronous orbit using radio frequency wireless power transmission.

Why have Solar Power Satellites not yet been built?

The key issues that prevented development centered around the size of the program, its cost, safety of wireless energy transmission, and international implications. These issues were compounded by the lack of the infrastructure required to support the program and insufficient validation of cost competitiveness with other sources. Also, it is a high technology space program that is outside the framework of the conservative electric utility industry.

Solar power satellites are only cost effective if implemented on a large scale. Geo-synchronous orbit must be used in order to maximize the sun exposure and maintain continuous energy availability. The transmitter size is dictated by the distance from the earth and the frequency of the power beam. The earth based rectenna also must be large to maximize capture of the beam energy. Given that the system must be implemented on a large scale, the cost of space transportation and the required space based infrastructure becomes the dominating development cost. Development cost of space transportation is driven by the need to dramatically lower the cost of space launches which can only be reduced to low enough levels by the use of fully reusable heavy lift launch vehicles which do not exist today.

The existing space transportation market has not been large enough to justify the huge development cost of a reusable heavy lift launch vehicle system. However, solar power satellites would create a large enough market if the perceived risk of their commercial viability is reduced to an acceptable level for the commercial investment community. The commercial investment community has been unwilling to invest in a long term, high cost project of this magnitude. The recent failure of the Iridium global satellite communication system has underscored the potential risks with space based commercial systems.

The concept of wireless communications is highly accepted and used the world over. The concept of transmitting power is not. The perception is that the power cannot be transmitted safely to earth.

Why Should they be Developed in the United States now?

Energy demand continues to grow as our population expands. The electronic age is totally reliant on electric power and is creating a new need for electric power. Many areas of the nation are experiencing energy shortages and significantly increased costs. United States electricity use is projected to increase by 32% in the next twenty years while worldwide electric energy use will grow by 75% in the same period. Worldwide oil production is projected to peak in the 2010 to 2015 time period with a precipitous decrease after that due to depletion of world reserves. Natural gas prices in the United States have doubled in the last year as the demand has grown for gas fired electrical generation plants.

Global warming and the need for reduction of CO2 emissions calls for the replacement of fossil fuel power plants with renewable nonpolluting energy sources. Even with increased use of today's knowledge of renewable energy sources carbon emissions are expected to rise 62% worldwide by 2020. If we have any hope for a reversal of global warming we must dramatically reduce our use of fossil fuels.

Solar power satellite development would reduce and eventually eliminate United States dependence on foreign oil imports. They would help reduce the international trade imbalance. Electric energy from solar power satellites can be delivered to any nation on the earth. The United States could become a major energy exporter. The market for electric energy will be enormous. Most important of all is the fact that whatever nation develops and controls the next major energy source will dominate the economy of the world.

In addition there are many potential spin-offs. These include:

* Generation of space tourism. The need to develop low-cost reusable space transports to deploy solar power satellites will open space to the vast economic potential of space tourism.

* Utilize solar power to manufacture rocket fuel on orbit from water for manned planetary missions.

* Provide large quantities of electric power on orbit for military applications.

* Provide large quantities of electric power to thrust vehicles into inter-planetary space.

* Open large-scale commercial access to space. The potential of space industrial parks could become a reality.

* Make the United States the preferred launch provider for the world.

The Situation Today

The situation is much different now than it was in 1980 when the earlier studies were terminated. In the ensuing years much has changed. Other programs have sponsored research and development of several of the enabling technologies and much of the required infrastructure has been developed. Studies have continued in several countries outside of the United States and some limited activity is sustained by individuals and companies on their own funds within the United States.

The development of terrestrial solar cells has caused the photovoltaic industry to grow from a very small specialty group of companies manufacturing expensive solar cells in laboratory quantities to an industry that is approaching maturity. Annual production is now well over a hundred megawatts and growing rapidly. Production processes have become automated and the number of different types of cells is greatly expanded. Thin film cells with efficiencies over 18% on metal film substrates and with inherent resistance to space radiation degradation will soon be in production. These cells will produce 1400 watts per kilogram of mass with a cost potential of 35 cents per watt. The decreased weight and cost will significantly reduce satellite cost and weight from earlier estimates.

Microwave oven magnetrons, manufactured by the tens of millions, have been converted into high-gain, phase-locked, amplifiers and shown they can be used to operate at high efficiency and at low noise levels in a wireless energy transmitter. Their low unit output eliminates the need for active cooling, further reducing system complexity.

Even though the Space Shuttle has not achieved its original goal of low-cost space transportation, it has proven the concept of reusability with aerodynamic reentry and landing. It and the Russian Mir Space Station are developing the knowledge base for manned operations in space. The International Space Station will greatly increase this base and is one of the key space infrastructure elements needed to develop solar power satellites.

As a result of all the developments that have taken place over the years since the 1970s it is now possible to consider another approach for solar power satellites. Most of the estimated development costs for the 1980 DOE/NASA reference design are no longer applicable. Over two-thirds of the total estimate was for infrastructure that is now being developed for other programs. Even though the low-cost and large payload capabilities necessary for space transportation of a space solar power system have not yet evolved, there is progress being made through commercial launch vehicles for communication satellites and the NASA/industry X33 program. One-third of the original 1980 cost estimates was to develop and build the first full-sized 5,000 megawatts output demonstration satellite. Based on the survey of several large utilities made by Solar Space Industries in 1994, a more realistic size for the first satellite is 1,000 to 2,000 megawatts output. This is the nominal size power plant a typical utility grid can handle without major problems. It also reduces the rectenna size of the reference system and therefore dramatically increases the potential receiver site locations.

With these considerations in mind it is now possible to take a fresh look at how to go about developing solar power satellites, lay out a development schedule, and identify who should be involved and from where the necessary funds will come.

This program is different from developing other potential energy sources. No research is required to develop the energy source for solar power satellites. It already exists. The sun is a full scale, stable, long-life fusion reactor, located at a safe distance. All that is required is to design and build a conversion system that can operate in the benign environment of space. The basic technologies are all known and proven. It is primarily an engineering application task to integrate these technologies into a operational system rather than a scientific invention/research task.

An inherent feature of solar power satellites is their location in space outside the borders of any individual nation with their energy delivered to the earth by way of some form of wireless power transmission that must be compatible with other uses of the radio frequency spectrum. They must also be transported to space. Government involvement to coordinate international agreements covering frequency assignments, satellite locations, space traffic control and many other features of space operations is mandatory in order to prevent international conflicts. Solar power satellites will ultimately become part of the commercial electric utility industry and as such,

that industry could be expected to shoulder the majority of the burden of development. However, the utility industry is not the only one that will benefit from the development of solar power satellites. All of the people of the world will eventually be the benefactors, through reduced atmospheric pollution and the availability of ample energy in the future. As a result it makes sense that the development of solar power satellites be accomplished through a partnership of industries and governments of all the nations that wish to participate.

What the Government should do NOW to Initiate the development of Solar Power Satellites

In a partnership of US government and industry it is vital that the leadership and responsibilities of the various elements be clearly defined in order to prevent chaos. There are some logical parameters to outline how this can be done. The first step is to establish a lead nation. The United States is the logical leader in this area because of the breadth of technology infrastructure and capability that already exists, as well as the magnitude of financial resources available in its industry and financial community.

The primary role of the government in this partnership will be to provide leadership and seed money to initiate the program, coordinate international agreements, support the development of high technology multi-use infrastructure, and assume the risk of buying the first operational satellite. The United States Department of Energy is the responsible government agency in the USA. They need to form a Solar Power Satellite Program Office to coordinate international cooperation and to be the focal point for other participating US agencies such as NASA, the Environmental Protection Agency, Federal Communications Commission, State Department, Department of Defense and the Department of Commerce. NASA, because of its expertise in developing space technology, will have the biggest role and is the appropriate agency to support the development of the multi-use space technology and infrastructure.

The last element of the government role should be the purchase of the first operational satellite. A government owned utility such as Bonneville Power Administration is the logical buyer of the first unit. Bonneville with more than 20,000 megawatts of generating capacity and an extensive distribution system is large enough to absorb the power from a 1,000 to 2,000 megawatt power plant. In addition, there are sites within their service area where the rectenna could be built. The cost will be repaid by the revenue generated by the satellite. The main reason a government utility should buy the first unit is so the government would accept the initial financial risk.

The other half of the partnership is industry. Industry can provide most of the developmental funding and be responsible for the design and development of the system. It is essential that the satellites and the space transportation system be developed in a commercial environment if they are to be viable commercial ventures.

The government needs to take the initial steps that will make it possible for commercial development to take place. Government agencies should not attempt to design and develop a commercial system, rather their role should be to create the opportunity and incentives to provide for commercial development. The following tasks should be initiated immediately.

1) Fund a Ground Test Program to demonstrate the satellite functions of power generation, wireless power transmission system, and integration of the energy into a utility grid on the ground. The Ground Test Program could also demonstrate the capability of the relay satellite power transmission by simply introducing a reflector into the power transmission beam. Thus the same program can demonstrate both concepts. The funding requirement for this program is very modest. A comprehensive Ground Test Program could be conducted for $30 million a year for a period of three years. Much of it could be obtained by focusing the funds that are currently being con-

sidered for the Solar Power Satellite Program on the Ground Test Program. This program would demonstrate to the commercial world the technical capability, efficiency, and subsystem costs of the power generation and wireless power transmission portion of the system. With this demonstration in hand the commercial companies would have the evidence they need to justify commercial investment in the operational system. A description of the proposed Ground Test Program is attached as Appendix A to this statement.

2) Obtain frequency allocation for worldwide wireless power transmission for operational satellite systems. This is a crucial step needed at this time as the communications industry continues to search for additional frequencies. It is imperative that wireless power transmission establish its own frequency base. This should include 2.45 and 5.8 gigahertz as the absolute minimum.

3) Implement the commercial space development tax incentives currently being considered in Congress. The Zero Gravity, Zero Tax bill is particularly important to commercial development of space.

4) Incorporate space infrastructure development and tests for solar power satellites into the plans for the International Space Station. Funding could be obtained by modifying the test and operational plans. It would give commercial purpose for the International Space Station. A candidate list of potential tests to support Solar Power Satellite development and space infrastructure development is shown in the attached Appendix B.

5) Continue technology development for reusable space transportation systems.

6) Consider the implementation of loan guarantees for commercial development of reusable space transportation systems and other required space infrastructure systems.

7) Commit to the purchase of the first operational Solar Power Satellite after the successful completion of the Ground Test Program.

Appendix A

Ground Test Program Description
Objective:

Demonstrate the complete function of solar power satellites as an electric power generating system for the 21st Century and the function of an energy relay satellite. This would include verifying technology and cost viability of the system elements associated with power generation, transmission, power conversion, and integration into an electric utility grid. It would provide the required data to update the design of a full-scale solar power satellite system.

Give the electric power industry confidence in the soundness of the concept.

Concept:

The concept of the Ground Test Program is to build a small-scale solar cell array, (in the range of 50 to 250 kilowatts peak output); couple it to a phased array wireless power transmitter which would transmit the energy over a short distance (1 to 5 kilometers) to a receiving antenna (rectenna); that feeds the DC power output through an inverter/power controller into a commercial AC utility power grid. This is illustrated in the following figure.

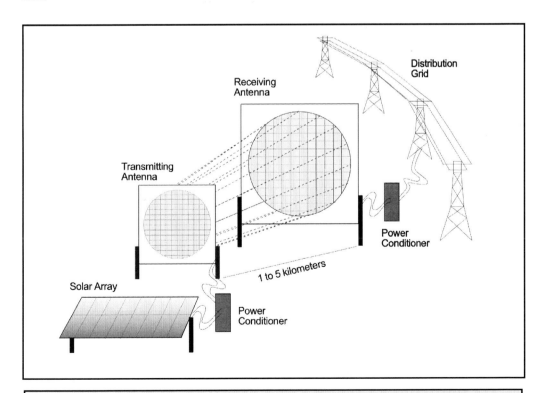

Figure 5: COMPONENTS OF THE GROUND TEST PROGRAM INCLUDE ALL PARTS OF THE SOLAR POWER SATELLITE SYSTEM

Each element of the system would be designed to incorporate several different technology approaches to be tested in the complete end to end test installation. For example the array could be made up of several 20 to 50 kilowatt sub-arrays of different types of cells, each with its own wiring scheme and power controllers. The transmitting antenna could have several types of radio frequency generators or have all of one type for one test and then be modified to another type for the next test. Different control circuitry could be tested to find the best approach for beam control. Various receiving antenna designs would be tested with associated power controllers integrated into the operation to test different designs for connecting into a commercial electrical grid.

The installation would duplicate all aspects of the power generating systems for the Solar Power Satellite concept, except for the space environment, and the range and size of the energy beam. The other functions of the satellite system have similar requirements to those associated with current communication satellites, except for size and the requirement to be assembled in space. These issues can be separated from the power generation function and verified by testing done on the International Space Station and Space Shuttle.

The concept of a relay satellite would be tested by changing the pointing direction of the transmitter and installing a radio frequency reflector that would reflect the transmitted energy onto the receiver. The radio frequency reflector acts in the same way as a mirror does for light.

Appendix B

International Space Station testing to support development of Solar Power Satellites

The International Space Station is one of the key infrastructure elements needed for the development of solar power satellites. The basic technology for the power generation and transmission will be developed and validated by a Ground Test Program, but this program does not address the issues unique to the space environment. These can only be tested in space. The Space Station is ideally suited to this task. Solar Power Satellites is a commercial program that will provide very large economic returns for the investment and by using the Space Station as the in-space test base will give the Space Station a commercial base to pay for its cost of operation.

A preliminary list of the research and development tasks and tests required for the development of solar power satellites, that could utilize the unique capabilities of the International Space Station, is shown in the following:

1) Test alternative structural concepts and assembly techniques for satellite structure.

2) Evaluate capability of alternative robotic assemble concepts.

3) Test radio-frequency generators and their characteristics when operating in the space environment.

4) Assembly techniques for the wireless power transmitter structure and sub-arrays.

5) Wireless energy beam formation and steering in the space environment.

6) Wireless power transmission in space from point to point (short range initially).

7) Wireless power transmission from the space station to the earth using the power capability of the space station. Evaluate beam formation and steering. Determine atmospheric effects and losses, including weather effects.

8) Evaluate and test potential candidate solar cells for performance in the space environment.

9) Test candidate mounting and assembly techniques for solar cells.

10) Tests of transmitting antenna rotary joint concepts and performance in the space environment.

11) Test ion thrusters and other electric thruster concepts for the attitude and station keeping control system. Determine characteristics and performance in the space environment and their compatibility with the solar array.

• • • • • • • • • • • •

With this plan commercial industry would have enough confidence to proceed with development of Solar Power Satellites. However, the result of my testimony was ————————nothing happened.

America had a crisis of leadership.

10   What Others Think

In the last Chapter I laid out the plan I proposed in my September 2000 testimony. It is still a sound plan, but there are other ideas and suggestions on what needs to be done today in our nation's space program to allow us to go forward with both space exploration and commercial development of space in our current economic crisis and what it can do for us. On February 20, 2009 the **Aerospace Technology Working Group** released a paper that explores the problems that our space program and NASA has suffered since the Apollo era. It identifies a probable cause of these problems, and gives recommendations on what needs to be done by the new Obama administration to rectify the mistakes of previous administrations and move us into a new era of economic growth that includes space based solar power. I am going to include it in its entirety in this chapter.

 Aerospace Technology Working Group

Aerospace Technology Working Group

Sustainable Space Exploration and Space Development A United Strategic Vision.

Feng Hsu, Ph. D.
NASA GSFC
Senior Fellow, Aerospace Technology Working Group

Ken Cox, Ph. D.
Founder & Director
Aerospace Technology Working Group

February 20, 2009

The general views expressed in this paper only represent the views of the individual authors, and they were not necessarily the views of organizations affiliated with each individual author. Questions and inquiries can be directed to Dr. F. Hsu, who is the author fully responsible for the contents of the document.

Copyright © F. Hsu, K. Cox 2009, All Rights Reserved

THIS MATERIAL MAY BE QUOTED OR REPRODUCED WITHOUT PRIOR PERMISSION,
PROVIDED APPROPRIATE CREDIT IS GIVEN TO THE AUTHOR AND THE AEROSPACE TECHNOLOGY WORKING GROUP

ABSTRACT

This paper presents and recommends a strategic and Unified Space Vision (USV) for comprehensive human space exploration and space development endeavors in the 21st century, through extensive analysis of complex space policy issues, to the new U.S. Administration, the NASA transition team and the broad domestic and international community. The proposed USV is a new paradigm of space policy that aims to rectify or replace the current Vision for Space Exploration (VSE), including its implementation plan, which has been pursued via the NASA Constellation program since its announcement by the former president in early 2004. We strongly believe that if adequately adopted, the USV should serve the long-term economic, diplomatic and exploration interests of this nation and others around the globe.

Introduction

In this chapter, the authors attempt to express some fresh thoughts or viewpoints, through honest, in-depth analysis of and discussions on a wide range of critical issues relating to strategic vision, national and international policy about human space exploration, and space-based economic development. The topics and general views presented in this chapter originated from a series of interviews and topic discussions between Dr. Buzz Aldrin, the honorary author, Dr. F. Hsu, the co-author, who represents the Aerospace Technology Working Group (ATWG) and is the responsible author of the manuscript, and Dr. Ken Cox the director of ATWG. The purpose of the interviews and topic discussions between the authors and members of the ATWG was to achieve consensus, through rigorous and candid debate on many critical strategic and space policy issues, to delineate a comprehensive and strategic vision on key strategy and policies pertaining to human space endeavors for consideration by the new U.S. administration. The space policy issues discussed and analyzed herein are based on a thorough assessment, not only looking at the current status and future perspectives of the U.S. space enterprise, but more importantly, through in-depth probing into the *organizational and political causes beneath the lessons drawn from the major successes and failures of NASA programs in its 5 decades of history.*

1. Key issues addressed in this chapter include:

* Why could NASA—in its present form of political and organizational paralysis—hardly deliver the kind of profound success achieved in the Apollo era?

* Why does the current Constellation program under the Bush VSE already have major schedule delays and cost overruns, and is increasingly at risk of repeating the past unsuccessful experience of NASA, such as the Space Shuttle program?

* Why is there an urgent need for realignment of America's space policy related to the proposed reform of NASA, its international policy and political governance?

* In particular, why does there need to be the establishment of a cabinet-level government entity, such as a "White House Advisory Council on Space" that serves as the overall government space authority, before evolving into a formal "Department of Space" (DOS) cabinet?

* Why is such a White House Advisory Council on Space (or a DOS) of paramount importance for providing leadership on all strategic space policy directions, overseeing NASA on Space

Exploration programs, supporting industry and national security space needs, incubating private space technology sectors, and devoting its main focus on promoting and leading the nation's effort for space development or space-based human economic and commercial infrastructure advancement?

* We firmly believe that a new paradigm of the strategic and unified space vision (USV) and related policies recommended herein, which have been drawn from candid and extensive discussions on above issues with members of ATWG and across the space community, will serve the long-term economic, diplomatic and exploration interests of this nation and others around the globe.

2. The Need of a New Vision for Transforming America into a Space-faring Nation

There have been heated debates in the public as well as within the space-science, industry and technology communities regarding the wisdom of the current Vision for Space Exploration (VSE), and its proposed implementation, as crafted and set out by the previous administration. More than 5 years have now gone by since its announcement in early 2004. It has become increasingly apparent that the thoughts and rationale that went into the formulation of the existing VSE and its implementation were quite problematic, and perhaps even lacked a strong strategic merit, to say the least. In fact, many of us in the space and intellectual communities find the VSE's lack of strategic vigor not much of a surprise, especially considering the rudimentary decision-making apparatus and processes of the previous administration, which led to many other lackluster major decisions on national and international imperatives.

In our view, there were several fundamental problems with the Bush Vision and its implementation for Space Exploration inherited from the get-go:

(1) Due to the lack of well-informed debate, engaging a broad range of the space and science community, policymakers, and the general public, the Bush VSE was crafted without the thorough reviews and studies necessary at the strategic space policy level. And in particular, such an almost-Apollo-style, huge national program of long-lasting impacts on national resources and sustainable space development was imposed to the American people, without learning the lessons of major program failures, successes, and key performance history of NASA since the Apollo era.

(2) The VSE lacks strategic merit, which can only be built upon a sufficiently vetted decision-making process of logic and analytic rigor. Especially, such process should have been scrutinized through hearings to engage the American public and politicians. Instead, the Bush VSE was a product of a blind and near-childish emotional response to a series of domestic and international geopolitical events that occurred in 2003, such as the launch of China's Shenzou-5 manned spacecraft on the 15th of October and the STS-107 (Columbia) Space Shuttle disaster in early February.

(3) Most notably, the political motives behind the sudden announcement of VSE by the Bush White House were severely undermined by the fact that the American public and politicians alike were largely distressed by the then chaotic situation of the war in Iraq, in which our nation and the executive branch were confronted with huge financial and political burdens from the two ongoing and costly wars in the Middle East.

(4). The budget necessary to fulfill Bush's VSE and the planned implementation for Space Exploration has far exceeded any financial resources available to this nation, as indicated by a recent GAO report. Therefore, many escalated budget cuts to earth monitoring, space science and robotic exploration programs may be inevitable in order to compensate for the extremely costly Constellation program, which was sold to congress in a hurry, with such unbelievable timing.

(5) The VSE falls short of addressing the national and international needs of human endeavors for space development objectives. Especially, the Bush VSE missed (or lacked) almost entirely any strategic vision and goals

for supporting and enabling space-based human economic expansion or industrialization in space. Such critical aspects of human space activities are fundamentally unique, and are quite different kinds of challenges from the space exploration activities undertaken by NASA.

(6) The current thrust of the Bush VSE to return humans to the Moon (and to build a costly lunar post without international participation and support) lacks political resonance. The American public and its political constituency in the U.S. congress are largely uninterested in supporting such a costly Apollo-all-over-again national program: "Been there, done that" rules apply. As a result, after receiving less than adequate funding from the Administration that proposed it, the Bush Vision for Space Exploration is unlikely to get more support from any new Administration, much less a chance of getting continued support from an Administration (like President Obama's) that is largely surrounded by visionaries and leaders with strategic and intellectual strength.

Undoubtedly, change must be brought about to rectify the Bush VSE and its implementation plan. And more importantly, a renewed or unified space vision—one addressing the comprehensive needs for sustainable space exploration and space development and that has high strategic merit, strong political support and financial affordability strategies from the American public—needs to be brought to light.

3. A Unified Vision for Concurrent Space Exploration and Space Development

We propose herein, and call for such a strategic and grand unified vision for both space exploration (VSE) and space development (VSD). This new unified space vision (USV) should be a comprehensive and balanced approach that addresses the long-term concurrent needs of space and science explorations, as well as the needs for space-based human economic development, which will benefit all of humanity while fostering world peace. It is a new paradigm of space vision with four critical strategic components:

(1) A vision of sustainable and affordable space exploration efforts that aims at probing for and discovery of unknown (or known) planetary destinations beyond the earth-moon system. Under this foresight, the space-access developments within LEO (low Earth orbit), including major elements of the constellation program from the Bush VSE, need to be regarded as space (economic) development activities to be achieved via international partnerships.

(2) A vision of space exploration that is fundamentally based on the concurrent development of an affordable LEO space-transportation infrastructure that will not only allow sustainable space exploration endeavors beyond low earth orbit, but more importantly, enable the rapid human

economic and commercial expansion into the Earth-moon system.

(3) Such a vision of space exploration and space-based economic development must be achieved and supported by the commercial and private sectors, along with broad international participation, using human collaborative endeavors in space as a strong catalyst or mutual bond for fostering world peace, thus collectively resolving humanity's profound energy and climate change challenges.

(4) Recognizing the distinctive strategic goals and objectives in both aspects of space exploration (VSE) and space development (VSD) activities. Therefore, restructuring and realignment of NASA's role is necessary for maximizing its potential in achieving technology R&D and space (and science) exploration objectives, whereas a separate government entity is needed in promoting economic-infrastructure achievement in space.

This is a grand strategic vision for concurrent or unified space exploration and space development goals and objectives, one in which space and science exploration endeavors can be largely funded, embraced and sustained by tapping into the financial and international resources through human economic, technology and commercial development in the LEO environment. To fulfill such a unified space vision, America must maintain its leadership in the implementation of such a strategic space vision through sharing space development and infrastructure build-up responsibilities, sharing technological and financial resources. That will allow for fostering a culture of shared responsibility with all international partners around the globe by utilization of human collective intelligence.

Mankind has achieved horizontal exploration and economic expansion around our own planet in the recent 500 years of human history. It is now time, at the dawn of the 21st century, for humanity to embark on vertical and outward expansion into space—not only for explorations to challenging planetary destinations, but more importantly, for revolutionary space-based economic and commercial development and space industrialization to the high frontiers. It is therefore imperative that we lead the march for transforming U.S. into a truly space-faring nation, and for all of humanity to become a space-faring civilization in order to prosper peacefully and forever survive in this planet and beyond!

4. Space Development vs. Space Exploration—What NASA Can or Cannot Deliver

The U.S. space program and NASA have had astonishing successes back in the Apollo moon-landing era nearly four decades ago. Unfortunately, such astonishing success were replaced rather quickly by frustrations and a series of unsuccessful programs, severe cost overruns, project cancellations, and decimated achievement of program goals, as represented by the costly and risky Space Shuttle program, along with many other short-lived programs ever since the early 1970s. Most of us within the space technical community would understand that development of affordable access to space, or low-cost space transportation capability, is the stepping stone not only for humanity to expand our economic presence in low earth orbit, but it also allows us to explore new destinations beyond LEO in a sustainable manner. However, many in the public, some even within the space advocacy community, find it hard to comprehend why such a change of NASA's fortune or track record occurred so dramatically—from the profound successes of the Apollo project to the devastating failure and disastrous and wasteful fate of the Space Shuttle program.

Frankly, we believe that the causes of such a tragic phenomenon were rooted deeply in the very reasons that led to the creation of such a U.S. space agency called NASA back in the late 1950s. In other words, the change that occurred in NASA, as indicated in its track record, was inevitable and destined to happen, due to the very nature of its organizational or political governance, which was mainly created to respond to the challenges of the space race during the cold war era. Indeed, the lackluster performance of post-Apollo NASA was predetermined nearly from the day it was created. Yes, it was a unique organization extremely capable of taking on urgent national challenges like the Apollo, and in winning the space race, NASA did exactly what it was supposed to achieve!

It would be hard to believe that NASA could fail to achieve success of the Apollo project, given the fact that it enjoyed full public, political and budgetary support throughout the entire project, and that the NASA budget back then was more than 10% of the U.S. GDP, a factor of 5 or greater than the post-Apollo era. Furthermore, such solid societal support of NASA had never blinked, even under circumstances such that our first Apollo mission was launched eight months after the Apollo-1 fire disaster, and that Apollo-11 was launched some three months after the near-miss anomaly of Apollo 8—a profound spirit that America had back then that has since been lost for taking risks and constantly pushing the envelope of engineering and science in the new frontiers!

It was evident that NASA's fortune of managing successful programs took a sudden dive for the worse right after the Apollo-17 mission was completed in 1972. Some might believe the obscure budget level and weak political support of NASA was to blame for the downturn of the agency in the post-Apollo era. However, while there was some skin-deep truth to such blame, the real truth lies deeply beneath the geopolitical complexities of the inherent NASA governing paradigm under which the agency was created: a military entity converted for the purpose of winning the space race. In other words, the space agency was much like a child born prematurely out of a C-section to satisfy tactical needs during the space race, rather than a well-structured government institution created for meeting America's strategic goals and its national or international interests. This was something of uttermost strategic importance that had never thought about or debated back then, due to historic reasons, and we owe it to the nation, the American people and all of humanity to do the needed rethinking, and therefore reforming of NASA at this pivotal crossroad!

America's national strategic goal or interests should include strengthening the U.S. and world economy, enhancing our leadership in science exploration and technology development, ensuring national security, and promoting world peace for sustainable human development. We need to understand that our political systems and the general public within our democratic society tend to respond to urgent needs very decisively and effectively when competing against an external threat, but they perform poorly during peace time, when there are no apparent or imminent external threats or perceived enemies. Yet, often, such "urgent needs" of national interests are either short-term objectives or responsive tactical goals without much of strategic value. In fact, the several major U.S. national labs created under the DOE are excellent examples of how government institutions could be adequately established and transformed to deliver strategic value in terms of science and technology advancements for the nation.

Clearly, due to the space race, the Apollo program started without much strategic vision or planning from the get-go, so America's space program was destined to lose direction soon after winning the space race. And this explains why the hardware and launch & crew vehicle systems were created, either without any strategic values, or they simply lost any such long-term applications (if there were any) immediately after the completion of the program. In fact, the lack of documentation and well-managed institutional and corporate memories on critical technologies from the Apollo project speaks for itself. Because of this lack of memory and well-kept technical heritages within the space agency, NASA and its current Constellation program has experienced great difficulty in the past few years in trying to understand and benefit from some critical technical achievements (such as the Saturn-V launch vehicle systems, etc.) that were successfully dealt with back in the Apollo project.

As a nation striving to prosper and build our financial and technological strength during today's post-cold-war times and under an increasingly globalized and mutually dependent world economy, America cannot (and must not) afford another such huge spending space exploration program—one that might end up winning (or even provoking) an unwanted space race, or win-

ning tactical space goals (such as beating others in building a costly lunar post first), but ultimately end up failing the nation in skyrocketing debt, or hurting America's long-term interests from wasteful space programs of little strategic and economic values.

Obviously, given the existing status quo and the political governance paradigm of the U.S. space agency that evolved from the space race era, what NASA cannot deliver are any successful, affordable or sustainable space programs (especially any space transportation or infrastructure development programs) without severe budget overruns or schedule delays during a post-cold war international geopolitical environment. That is why under such a systemic paralysis of NASA, huge cost overruns or schedule delays have become the rule rather than the exception. That is also why some of us within the space community have been either reluctant to accept or very much against any ideas of a broader international collaborations strategy in human space exploration or development endeavors.

Furthermore, with the existing NASA structure, we tend to amplify or exaggerate the threats of any space achievements from other nations, such as when the launch of China's Shenzhou-5 was perceived as a big threat, and became the primary drive behind the quick sale of the Bush VSE. So the lack of strategic vigor is almost destined to repeat the failures of the agency and thus ultimately hurt our

strategic interests. Has anyone imagined why we needed to even respond to a "space race to the Moon" when America won the race four decades ago?! Has anyone imagined what the Bush's VSE or the Constellation architecture would look like back in late 2003 had the Changhe-1 of China been designed to be a Mars Lander, instead of a lunar orbiter? We believe that a responsive vision for space exploration with independent wisdom and strategic merit is good for America, but a responsive space vision largely influenced or misled by external events without strategic merit can be detrimental, not only to America's long-term interests, but the interests of all humanity as well.

The inability of NASA to efficiently handle space exploration programs aimed at peaceful discoveries is precisely because during the "peacetime" environment, all the needed budgetary and political support to sustain the operations of the ten-plus NASA centers and organizations was largely diminished, and replaced with the complex issues of fierce competing battles to survive (fund) within NASA organizations. Given such organizational deficiencies of NASA hierarchies as it was created, it simply became an organization incapable of planning or doing anything with bold strategic vision or value. The post-Apollo NASA, as it has been for the past 40 years, simply became a visionless jobs-providing enterprise that achieves little or nothing in space infrastructure development, especially in the effort for reusable or affordable launch-systems development.

Indeed, this explains why in the past, numerous NASA programs or proposals with excellent design concepts were discarded, whereas wasteful projects with costly or unnecessarily complex and risky designs survived. To name just a few: the Shuttle II, Shuttle-C, and the National Launch Systems (NLS) were replaced by the current architecture of the Space Shuttle; and the X-30, National Space Plan (NASP), the X-33 Single Stage to Orbit (SSTO), the Space Launch Initiative (SLI), and OSP (Orbital Space Plane) were cancelled one after another.

It is quite clear that without adequate reform and without long-overdue overhaul of NASA's organizational and political governing infrastructures right after the completion of Apollo project, NASA was destined to deliver wasteful or unsuccessful programs. In fact, the selection of the Space Shuttle system design concept and configuration was heavily influenced by NASA's internal and inter-state politics, a textbook case of such organizational deficiencies. That's why we believe that the current Constellation program is at high risk of repeating the post-Apollo "track record" of NASA and we strongly urge the reform of NASA and recommend, as elaborated in the following two sections, how the nation's space enterprise should be adequately chartered, managed, and guided by the proposed USV, to take on the concurrent challenges of fulfilling the strategic goals of space (and science) exploration and space development.

5. A Critical Path for Achieving the Unified Space Vision (USV)

We believe that NASA, as a space agency of the U.S. government, is an adequate and rightful government institution for conducting the nation's space exploration programs and projects, which primarily aim at explorations, including earth- and space-science discoveries and a planetary defense effort. However, as discussed earlier, NASA does need serious reform or significant organizational overhaul, with respect to institutional governance and enterprise structures, to become capable of conducting successful large-scale space exploration programs and projects.

Particularly, NASA needs reforms on the fund distribution mechanism and budgetary approach for effective governance and operation in this post-cold-war and globalized modern era, in which there is no strategic need for engaging in any type of Apollo-like space-race programs. Although an efficient and functioning NASA is critical to conduct the nation's space exploration programs successfully, NASA and its space exploration (manned or robotic planetary science) effort can represent only part of the large picture of America's needed human activities in space. In other words, there is a much broader category of human *space activities (referred in this chapter as the VSD, or Space Development category) that cannot be handled or managed effectively or successfully by such a government agency as NASA.*

In our view, even with adequate reform in its governance model, NASA is not a rightful institution to lead or manage the nation's business in Space Development projects. This is because human space development activities, such as development of affordable launch vehicles, RLVs, space-based solar power, space touring capabilities, communication satellites, and trans-earth or trans-lunar space transportation infrastructure systems, are primarily human economic and commercial development endeavors that are not only cost-benefit-sensitive in project management, but are in the nature of business activities and are thus subject to fundamental business principles related to profitability, sustainability, and market development, etc. Whereas, in space and science exploration, by its nature and definition, there are basic human scientific research and development (R&D) activities that require exploring the unknowns, pushing the envelope of new frontiers or taking higher risks with full government and public support, and these need to be invested in solely by taxpayer contributions.

Therefore, NASA by its very existence, like many U.S. national research laboratories, is supposed to be a logical R&D organization that should mainly dedicate itself to exploration,

planetary research, scientific discovery and technology development programs. Likewise, the proposed cabinet-level U.S. Department of Space (DOS), as discussed earlier, should manage and take charge of the government functions of supporting and incubating space-based industrial capability and transportation infrastructure development. Unlike NASA, the key role of the DOS should be to support and foster strong government-business partnerships (much like the current NASA COTS program, but at orders of magnitude increased scale) with space industry and the private sector to promote space infrastructure development. This directly benefits the national and world economy and brings investment returns to taxpayers, not just by creating more high-tech jobs, but also supporting NASA on more ambitious space exploration programs.

We firmly believe this is the strategic approach or Unified Space Vision (USV) that will ultimately achieve its goal of sustainable human space explorations and spread of mankind's presence beyond Earth. Again, the rationale behind such a proposed space policy (i.e., USV, a concurrent VSE & VSD approach with VSE largely supported and embraced by VSD, and the VSD goals and responsibilities largely separated from NASA and placed under the leadership of DOS) lies within the principles of organizational management theory. The programmatic principles or management culture applied in handling R&D projects (space exploration) are funda-

mentally different from the principles and organizational culture that are effective in managing space development programs, So NASA should not be too conservative in exploring new frontiers and unknowns by accepting just an adequate level of technical and programmatic risks. However, DOS in its space-development efforts should manage projects based on strict business, cost-benefit and market principles in order to develop products (such as high-reliability launch vehicles) that are affordable for commercial space applications, thus contributing directly to economic human expansion into the Earth-moon system.

Likewise, if a space exploration project like a manned Mars mission is managed by bean counters who fear taking even a moderate level of technical risks, this will not only make space exploration missions far too expensive to afford, but also render our likelihood of achieving any kind of exploration successes comparable to the Apollo project. Therefore, we suggest that such problematic management policies as full-cost accounting, or most ITAR restrictions being widely applied to NASA should be removed to enable the full potential of the space agency in its space and science exploration activities.

There are limited financial resources from the U.S. government, which is now struggling with unprecedented high budget deficit and is confronted with extremely costly ongoing wars. So it is nearly irresponsible to impose on the nation and its people an Apollo-like, huge spending lunar-based space exploration program. There is neither significant (or short-term) science value nor space exploration and operation value in revisiting an earth-orbit destination that was explored by mankind four decades ago. Given today's decimated American economic condition, we must adapt a concurrent and comprehensive space exploration and space development strategy that is not only affordable but can be mutually supported.

In particular, we recommend adapting a strategic Unified Space Vision (USV) by which space and science exploration goals (VSE) are to be achieved, funded and compensated not only by limited government investment, but more importantly, supported by promoting space development projects that enable low-cost space transportation capabilities. Such affordable space capabilities can only be achieved by encouraging extensive private and entrepreneurial investment and government-industrial partnerships for space-based commercial and economic infrastructure construction and industrialization (as proposed in VSD). Some detailed and key elements of space exploration activities, within the framework of USV, are envisioned and can be achieved through the following recommended critical path for affordable and sustainable space endeavors:

(1)The U.S. should adopt a renewed vision for space exploration

(VSE) that aims at returning the U.S. to the forefront in space and leading humanity's space exploration challenges to new frontiers, rather than repeating what the nation and mankind did with the original moon landings. Under this vision, we recommend reform of NASA, the establishment of a White House Council on Space (or DOS), and adoption of a strategic and unified vision for a comprehensive and concurrent effort in space (implementing both VSE & VSD) for the nation's space endeavors. In this strategy, we propose that the current NASA effort of returning to the moon should be regarded as part of human Space Development, to be implemented by DOS, which is a key element of overall space transportation infrastructure development activities for human economic and commercial expansion into the Earth-Moon orbit systems.

(2) We must adopt a unified strategic vision and related space policy in both space exploration and space-based economic development activities for engaging the international community and all industrial and private sectors by providing leadership and assistance to other nations, especially in developing countries, for lunar exploration missions, thus establishing an international presence for lunar & planetary science explorations, and conduct necessary space technology tests for risk reduction of a manned mission beyond the earth-moon system. Major investment on lunar transportation and surface systems development should be based on international

resources, and any U.S. investment in lunar missions should be assessed by its necessity and technical relevance and risk reduction benefits of manned Mars explorations, or any other manned explorations beyond the Earth-Moon confinement.

(3) A new paradigm or change of mindset in international collaboration of space explorations must be adopted on the part of the U.S. government and its general public. We must educate the public and our politicians to realize that human space exploration must be a global effort that is shared and supported by all of humanity. We should avoid, even a hint of arrogance and abandoning the old way of U.S.-led international space collaborations where we dictate all technical and programmatic outcomes. Most importantly, we must use space as a strategic tool of U.S. diplomacy not only to strengthen relations with our allies, but also for enhancing mutual understandings, diffusing and transforming confrontations with all other nations on earth, especially developing and non-democratic nations, with the ultimate goal of spreading democracy and the American democratic values. We must avoid provoking any new space race, as it has a high risk of getting everyone involved in a loss-loss combative cycle. The history of the U.S.-Japan and U.S. China relationships during and after WWII should shed light on our strategic thinking on this critical issue. Wars and confrontations could always be diffused and avoided, just as the former friendly and hostile nations were

transformed or altered throughout American history, and outcomes all depend on collective human wisdom.

(4) The U.S. space exploration goal should focus primarily on exploring unknown and new destinations by use of robotic exploration as much as is practical. However, the new vision (or VSE) must be more of an interplanetary-exploring nature, with a manned mission to NEO or staging at, and returning from the sun-Earth L2 libration point, as preparation for a precursor mission to Mars' moon Phobos, followed by manned missions to land on Mars. To achieve these goals, the U.S. should develop a Deep Space Habitat (deep space experiment module or station beyond low Earth orbit), complete with artificially produced gravity, for use in flying to destinations or to reside at various libration points (such as the moon-Earth L1 or sun-Earth L2 staging points), or to orbit various NEO destinations. This experiment module or habitat could be used as part of the "fly-by" and orbit program mentioned above.

(5) With the success of a manned NEO or L2 staging mission, a manned mission to Phobos can be carried out prior to a manned mission to Mars. Also, a one-way manned mission to Mars can be considered, with sufficient Mars crew Hub capacity and in situ resource utilization (ISRU) capabilities delivered prior to the arrival of the first manned Mars mission. We also recommend an R&D effort and demonstration projects on space-based solar power (SBSP, which offers a great potential for electric propulsion and power resources that can be utilized for deep space exploration missions. But more importantly, its key technology components can be shared or used by many other space applications, including future supply of base load power from space for terrestrial electrical energy demands.

(6) The above exploration goals (lead by NASA and the international community) can not be achieved unless a cost-effective HLV (heavy launch vehicle) or affordable LEO transportation infrastructure is developed first, or developed concurrently by DOS and its global collaborators. Such as low-cost crew LV (launch vehicle) and cargo HLV system development should be the task of highest U.S. short-term priority in space development, as they are not only crucial for supporting all strategic space exploration goals but also imperative for space-based economic and commercial development, such as development and demonstration of space based solar power and space tourist infrastructure system capabilities.

6. Propel Humanity's Outward Expansion into Space-based Economic Frontiers

As discussed in the previous sections, a space agency without reforms, as it still exists today, born out of the cold war era half a century ago, worked well for the space race, but is unlikely to deliver space-development achievements that benefit our national economy. It is also more likely to resist international participation, or even likely to exacerbate external threats and provoke an unnecessary or detrimental space race. What the U.S. and the international community urgently need in the 21 century, under a globalized world economy, for confronting the global climate change, energy and economic challenges is, however, opening the new frontier of economic and commercial development in space, especially industrialization in the Earth-moon LEO system. The recent history of the profound leap-forward of human economic development, triggered by the opening of commercial air transportation capability, must shed light on how we should embark on the next giant leap of humanity's economic and commercial expansion into low earth orbit.

Technology innovations have always lifted human society out of the economic gridlocks, and have led mankind from many of the worst economic crises to vast industrialization and enduring prosperity and growth. The history of human civilization has shown that technology innovations and human ingenuity are our best hope to power humanity out of any crisis, and especially a U.S.-lead human economic development into low earth orbit that will not only lift us out of the current acute global depression, but will most certainly bring about the next economic and industrial revolution beyond the confinement of Earth grav-

ity. Commercial aircraft transportation and operations in the past 100 years since the Wright Brothers' first successful test flight have advanced significantly in all areas, and have contributed tremendously to the world economy and modern civilization.

Nonetheless, space access capability and associated LEO infrastructure has generally not advanced in nearly half a century. Particularly, as elaborated in the previous sections, given the current plans under the Bush VSE for the next generation of human space transportation being pursued by NASA, there exists little hope of making any substantial improvements in safety, affordability, or commercial operations of any LEO transportation infrastructure for another generation.

With the impact of the upcoming termination of Space Shuttle operations, as guided by the Bush VSE, it is very clear that the U.S. needs substantially improved crew and cargo space access capabilities, and such improved space access capabilities are largely represented by a two-stage, fully reusable launch vehicle (RLV) system (in the short- to mid-term). An evolutionary infrastructure buildup of such a RLV system that is largely based on existing heritage or capabilities should be a key element of a reliable and low-cost cargo/crew space transportation development. Indeed, development and government investment in such an affordable space transportation infrastructure in the Earth-Moon system is of paramount importance; it's all about the crossroads the U.S. is at with the current economic crisis and how Space could be a key part of the answer.

A key component of a sound strategic space vision that was missed almost entirely by the Bush VSE is the vision for space development (VSD), or a space-based economic and commercial expansion into low earth orbit. Such a vision should be to place the highest priority on embarking on a national and international strategic space development goal that will ensure the technological, and with it, the economical leadership of America for the 21 century and the next few hundred years ahead. Otherwise, we risk continuing on the course of the Bush VSE, allowing it to drift into the back waters of history.

Investing in space infrastructure development—such as low-cost RLV systems or fully reusable, two-stage (or ultimately single-stage) space access system developed as an extension of safe and reliable airplane operations or investing in SBSP (space based solar power) and space tourism infrastructures as a significant part of the national space economy and energy programs—is the choice of a strategic space goal that certainly will re-ignite the American spirit and jump-start its high-tech manufacturing sector. It will send a profound message to the world: that America is still a nation where great bold endeavors are the order of the day. , Or else, it will be a message that we will allow the nation to contin-

ue its drift into obscurity and signal that America's greatest days are in the past.

Yes, there may be those who are against any space-based economic development, such as developing a low cost RLV capability, a stepping stone that could enable a whole host of private space industries, such as space tourism and space energy industries. Many of us may also argue that RLV or SBSP are too expensive or too hard to be realized. However, as Americans, we must not forgot what makes a nation and its people thrive and prosper are not based on what they do for easy or short-term gains; it's largely based on what the nation and its people do that most others dare not to do or cannot do!

This nation created the Manhattan project more than 60 years ago when we had only a rudimentary idea back then of how nuclear power could be tamed for electricity generation that can be competitive economically. We decided to take on the seemingly sci-fi Apollo project when we lacked definitive ideas of how to even get to the moon, much less knowing how to return humans safely. We sent a Lewis-Clark team on the expedition of the western frontier more than 200 years ago, against the fierce resistance of the bean-counters within the U.S. congress. We believe no one would argue the facts of the profound benefits this nation has received from these bold projects and of strategic investments in all economic, scientific and social *fronts. We did all such great human endeavors in the past not because it's easy, but precisely because it's hard!*

We recommend a new paradigm of a strategic vision for space development (VSD) be considered by the new administration, consisting of the following key strategic components, as a viable roadmap for propelling America and humanity's vertical expansion into space-based economic and commercial frontiers:

1) Set the goal of space transportation infrastructure development within the Earth-moon system as the highest priority by the new administration in its space policy, based on a strategic vision for space development (VSD), to be implemented by the proposed DOS. To be successful, the U.S. should build strong support and wide participation from the international community.

2) In this effort to achieve the proposed VSD, the U.S. and its international partners should focus heavily on the development of RLVs, (reusable launch vehicles) such as crew & cargo transport and launch vehicle systems with top-level requirements of low-cost, low system complexity, and aircraft-like reliability, maintainability and operability.

3) Develop and establish an international Fuel-Depot and Orbital Staging or Service point (station)

in the LEO environment that supports and services commercial space-transportation traffic needs or capabilities, such as space tourist flights, Lunar and earth orbital transfers, and commercial satellite services.

4) Promote and support the establishment and construction of Space port infrastructure development in several strategic locations within the U.S. and around the globe, which will be utilized to meet the emerging demand of increased commercial launch and space-transport economic activities.

5) Develop enabling space infrastructure and observation and tracking capabilities for planetary defense. In particular, develop ground and orbital systems, in close collaboration with international partners, for monitoring, tracking and deflecting any asteroids, comets, and other cosmic objects in near-earth orbit, which are at credible risk of threatening the safety of our home planet.

6) Invest in space-based solar power (SBSP) research and development, by first funding a series of space-to-space or space-to-earth SBSP demonstration projects. Technology demonstrations, such as wireless power transmission (WPT), high-efficiency microwave beam generation and control, system safety and reliability, on-orbit robotic assembly technology, and deployment of large-scale orbital solar structures will help reduce risks, thus triggering large-scale investments by private industries, and ultimately lead to great potentials of harnessing solar energy from space to alleviate our dependence on fossil fuels and mitigating global climate-change risks.

7. A New Space Economy with a Transformed Global Collaborative Paradigm

History has brought mankind to the brink of an unprecedented era of crisis and challenges. However, crisis and challenges encompass new opportunities for all of mankind, as implied in the Chinese word for "crisis," which also means "opportunity." Our crisis in the world economy, energy resources and global climate change are dire, but our opportunities for science, technology advancement and human economic expansion in space are enormous. Having evolved and survived on earth for millions of years, through constant struggle for change, we humans must once again, expand and adopt new economic spheres, and elevate from an Earth-bond civilization to a space-faring civilization in the face of crisis. Much like our ancestors learned to adapt using fire and tool-making skills, and evolved from primitive tribal-based societies into the collaborative agricultural civilization; from isolated regional economies to a globalized world economy. Now is the time for humanity to develop space and industrialize the Earth-Moon system, *making it a key part of global economic revitalization for a whole new sustainable and elevated human civilization.*

Many of us believe that mankind must solve all our crises on earth before expanding into space can be achieved successfully and peacefully. In fact, humanity isn't going to solve all its problems here on earth, ever. While resolving some of our crises, humanity always creates more. Regardless, mankind goes into space for reasons that our ancestors had historically gone elsewhere: for adventure with unknowns, resources, freedom, and better lives. The recent human history of industrial revolutions, along with the current collapses of the world's economy and energy and financial markets, has taught us a harsh lesson: that merely manipulating financial capital and producing services has failed to build a sustainable global economy for mankind. Instead of fighting over what's limited and restricting human development on this planet, we must now expand our horizons, and look upward and outward for resources, embarking on economic and commercial development into space.

Bold strategic vision supported by strong government and global leadership in technology and infrastructure development has always brought humanity out of our economic and political crises, much like the bold vision of the transcontinental railway systems supported by president Lincoln in the mist of the civil war crisis, or like the infrastructure buildup of the massive U.S. interstate highway systems called for by president Eisenhower in the aftermath of the great depression back in the 1930s, or like the government investment of Internet technology and information infrastructure buildups in the early 1990s supported by president Clinton. Now is the time, more than ever, for yet another bold vision and for America's strategic leadership to bring humanity out of our crisis by promoting and investing heavily on the final frontier of human development in space. Indeed, whether to make space industrialization an integral part of our strategy, and a key component of a stimulus for our economic recovery is all about the crossroads the U.S. and the rest of the world must decide on in the face of the many crises humanity has encountered.

Mankind, in the current stages of our single-planet civilization, may feel compelled or threatened to fight over resources and living space on the surface of the earth. However, such an inherent condition and competitive human psychology (deep in our consciousness) will most likely change by expanding the human horizon outward into space. As evidenced by human experience as astronauts, the "overview effect" will be the most profound nature bond for humanity to cherish one another, when we first looked back at our obscure blue home planet from deep space.

We must not underestimate the paramount importance of expanding human habitats outside the earth confinement as a critical benefit contributing to the acceleration of human conscious evolution, and hence bringing about transformed geopolitical governance, and ultimately leading to sustainable and peaceful human development back on earth. Much like a political vacuum existed in the New World some five centuries ago, which allowed early American settlers to experiment with more efficient and just forms of government, there is little doubt that humanity's expansion into space will help us develop healthier and more peaceful societies on earth.

A bold strategic vision and strong leadership always require one to think outside the conventional paradigm and learn from history. Both houses of the U.S. congress have been debating fiercely on the huge economic stimulus package proposed by the new president. But it's quite disappointing to see that only a small fraction (about 5 billion dollars) is allocated for the buildup of new infrastructures, such as a mass transit system for America. Much of the allocated investment is limited to repair of bridges and existing highway systems, without a hope, in our view, of seeing any major investment in space infrastructure buildups, such as affordable RLVs or other space transportation systems development.

What America desperately needs, beside short-term unemployment rescue measures, is to open up whole new commercial and economic frontiers, and with it, waves of technological and *industrial innovations. Space industrialization, alongside with renewable energy and mass transit infrastructure development, are key sectors of the emerging economic and technological revolutions that will lead humanity to the whole new realm of prosperity and sustainability. Unlike investment in a short-term stimulus project, the jobs and opportunities in space development that will benefit the U.S. and the world will be enormous, and will likely dwarf anything mankind has ever seen in the past. So we strongly urge the new administration and the U.S. congress to consider the strategic policies, as outlined below, to support the development of a transformative space-based global economy:*

a) Consider significant portion of the stimulus package to be allocated for supporting space development projects. In particular, supporting and incubating technologies and entrepreneurial partnerships of space projects for commercial orbital transportation

b) Support development of the space tourism industry, making it among the top priorities of space development projects. Help establish and enable the commercial tourist market place in zero-G, suborbital, and orbital environments.

c) Obtain a sizable portion of the funds by redirecting resources from the costly Ares launch vehicles and lunar base projects, to allow NASA to focus its resources and human capital on renewable energy R&D, including development of affordable solar, wind and geothermal energy systems and products, and cost-effective energy storage technologies. ISS, lunar base development and lunar science explorations should be the major focus on international collaborative programs.

d) Support earth and space science R&D projects to enhance earth and environment monitoring capabilities, and develop strategies and technologies to help mitigate and avert risks of natural disasters and catastrophic climate change, while protecting our natural environment.

e) Invest in space solar power research and demonstration projects, wireless power transmission and electric propulsion technologies, including related cost-effective and highly efficient electric engine systems applied to ground transportation vehicles, commercial airplanes, and space propulsion systems

Frankly, if we really wish to revitalize the U.S. economy and make it the most powerful world economic engine for centuries ahead, we must try not to put another 25 billion into saving the troubled auto industry, as it would only postpone the death of a key element in the obsolete U .S. economic base! Why not think about spending 10 billion on space infrastructure development, and another 15 billion in building the new mass transit infrastructure, such as high speed MLV transit network systems, which could change our economic model, which has reached its peak?! Has anyone seriously thought about the problems or sustainability of our car-based economic model, beyond the current fossil-fuel crisis? How much time do Americans need to waste on the roads every day before we start thinking outside the old economic realm?

Average Americans spend about one whole week time each year waiting at traffic lights, not to mention the horrible road jams, and collectively we waste a total of more than $78 billion per year sitting in cars without moving an inch. Clearly, the key for strategic recovery is to create new economic and business frontiers, and expand human presence and activities into space. Spending massive amount of borrowed money to "fix roads and bridges" may provide some short-term stimulus, but it may not be the best strategic idea for the future of America. Yes, it could put many folks to work temporarily, but it is not going to sustain them, simply because we are not creating anything new, but rather attempting to recover from the obsolete economic base by following the old tracks of a failed economy. Investments made in a mindset based on the existing economic paradigm will likely lead us to where we were before, and we will sooner or later find ourselves in the same traps.

Space industrialization is essential not only to the continued wellbeing of humanity on earth, but as a key step to assure the continued survival of the human species. We cannot continue to prosper and survive for long without tapping into the unlimited resources of our solar system. We urge the new president and the U.S. congress to support engaging broad international partners, provide U.S. leadership in both space development and space exploration endeavors, and promote human commercial and economic expansion into space, following the unified strategic space policy elaborated in this paper.

Finally, we envision with confidence, by the turn of the next century, that if we adequately implement such a bold and unified space vision (VSE/VSD), along with realizing other aspects of strategic goals of renewable energy development and environmental protection, the various earth orbit destinations, and luxury hotels and attractions on the Moon will be among the choices of travelers or tourists on Travelocity. Humanity will experience the profound reality of witnessing business executives or ordinary travelers boarding a space flight leaving New York or London in the early morning hours and arriving back home for dinner with families after several hours of meetings in Shanghai or Tokyo!

What better strategic vision can there be for the future of human space exploration and development than leading humanity on a solid track to becoming a space-faring civilization?

REFERENCES

Louis Friedman, Jacques Blamont, *A New Paradigm for a New Vision of Space*, The Planetary Society, Pasadena, CA. Nov. 2008.

William Claybaugh, Owen, K. Garriott, Michael Griffin et. al., *Extending Human Presence into the Solar System*, The Planetary Society, Pasadena, CA. July, 2004.

George Abbey, Neal Lane, John Muratore, *Maximizing NASA's Potential in Flight and on the Ground: Recommendations for the Next Administration*, James A. Baker III Institute for Public Policy, Rice University, January 20, 2009.

Feng Hsu, Romney Duffy, *Managing Risks in the Space Frontier, Beyond Earth—The Future of Humans in Space*, Apogee Books, 2006.

Buzz Aldrin, *Fly Me to L1*, The New York Times, 5. December 2003

Adriano Autino, et. al., *For a Politics of Support to Space*, The Space Renaissance Initiative, SRI, Nov., 2008.

Feng Hsu, *Harnessing the Sun— "Embarking on Humanity's Next Giant Leap"*, Proceedings of International Conference, Energy Challenges, Foundation For The Future, Seattle, March, 2007.

Jeff Krukin, *Space – A Place of Abundant Resources*, Space Daily, July 2004.

Acknowledgements

The authors are deeply indebted to many of our colleagues at ATWG for their tireless support and help in editing this paper. Special thanks are dedicated to Mr. Rick Elkcamp, Dr. Sherry Bell and Dr. Langdon Morris for their invaluable comments and suggestions in the edit of this manuscript. Our heartfelt thanks also go to Mr. Adriano Autino and Dr. Raymond D. Wright of the Space Renaissance Initiative for their insightful comments, which have helped make this paper a much better document, and to Amara D. Angelica and Annie Bynum, for copyediting this document.

• • • • • • • • • • • • •

As you can see others have observed the problems of our space program since the Apollo era. It is not surprising as NASA's charter is basically exploration and space technology. They are not in the business of supporting commercial development. However, when we look at history and consider the exploration, development, and exploitation cycle that all new discoveries go through it is not surprising. This cycle occurs in technology discoveries, exploration of our planet, exploring space, and new industrial discoveries to name just a few. I discussed this subject in **SUN POWER** and because of what has happened over the last few decades in space the following excerpt will show why the recommendation to establish a Department of Space as suggested by **ATWG** is so important.

From SUN POWER, Chapter 15, The New Frontier

As we look back in history, we find that humanity is always searching for a new frontier to explore and develop. If we do not find one, we become restless and try to take one from our neighbor, which often results in war.

And there are those who are not satisfied with the status quo, who have to sail beyond the horizon, climb the highest mountain, seek out the depths of the sea, search out the mystery of the atom or the magic of electrons running around inside a tiny chip of silicon. These are the explorers—whether of geography or of science—who are looking for new frontiers. Some are small and private; others change the course of the world. I believe it is worthwhile to examine some of these examples to see what might be happening to us today.

The Early Explorers

Columbus had a bizarre dream and convinced Queen Isabella that he could reach the Orient and all its great riches by sailing west. After she gave him his ships, he and his reluctant crew started out, much against the wisdom of the day that said they would fall off the edge of the world. Even though he found a new world, Columbus fell into disfavor and never did reap the benefits of his great discovery. Others followed to develop the land and to colonize—some to loot the riches, many to start new lives, others to seek freedom. They were the misfits, the mavericks, the restless, and the builders who developed new nations and reaped rich harvests of benefits undreamed of by Columbus. Eventually, in the colonies, it was the settlers who came and brought stability to the land where they would build their homes, businesses, farms, and eventually the greatest economic giant on earth.

Later, within the new nation called the United States of America, two explorers named Lewis and Clark convinced Congress to fund an expedition to the Pacific Northwest territory. Into unexplored wilderness they went, finding the glories of the Rockies, the Snake River, and the Columbia Gorge, returning with tales of wonder and excitement. In their paths followed the fur trappers, the traders, and rugged visionaries. These men could see beyond the wilderness to the day when settlers, industry, and commerce would thrive and the riches would flow. Flow they have, again far beyond the dreams of those early explorers and developers.

Wilbur and Orville Wright dreamed of flying like birds through the heavens—and they did. Once they proved it could be done, more followed, but in those early days it was difficult to imagine any way that flight could be of any practical use except for the personal thrill of flying. A few visionaries saw the day when the mail could be carried successfully, or even passengers. The military became the first to exploit the potential, however, and the skies of World War I were alive with the angry snarl of Spads and Fokkers amidst the rattle of machine guns. After the war, the airplane once again became a toy, but a much better toy. Gradually some of the early visions began to take on the semblance of reality. A few passengers were being carried and some of the mail was being delivered by air.

Then in 1927 Charles Lindberg electrified the world with his flight across the Atlantic, and aviation had come of age. Shortly thereafter, cabin stewardesses flew for the first time on the elegant three-engine Boeing 80A of the Boeing Air Transportation system (the forerunner of United Air Lines). Soon the Douglas DC-3 revolutionized air passenger service. Some even thought it was the ultimate aircraft, never to be exceeded.

The magnitude of today's commerce in the sky boggles the mind. At any given time, there are probably more than 3,000 commercial passenger and cargo aircraft in the air somewhere in the world. Even if they only average 100 passengers each flight, that represents more than 300,000 people cruising the skies of the world at any time. What amazing progress for a frontier opened less than a century ago to the skepticism of many highly educated people who pronounced it a lot of foolishness and waste.

The lessons to be learned here are typical of the evolutionary development cycle of all new frontiers, whether geographical or technical. This cycle begins with exploration, frequently the result of one person's dream. More often, it is the culmination of the efforts of several dedicated people striving to accomplish that which has never been done before. It is a period of excitement and thrills. Even the uninvolved are intrigued and watch in awe as spectators. They may even contribute in a monetary way. Once the goal has been reached, however, their interest wanes. The unknown is now known. It was fun while it was happening, but what good is it?

The next part of the cycle is much more difficult. This is the development phase. All the excitement of the initial exploration is gone. Very little glory is left to the second person who does anything. Not enough is known about the new frontier to be able to accomplish anything very useful, except to gain further experience and knowledge. It is hard to find anyone anxious to fund an activity whose objective is simply more experience and knowledge.

Now come the visionaries—those people who can see from the scant base of available knowledge into the future of practical application. They clearly see the benefits and know that an investment made in developing experience and knowledge will someday lead to large returns, even though the development period can be decades in length with only limited returns starting immediately. Sometimes military exploitation stimulates the development cycle as it did with the airplane.

As the development cycle progresses and the experience and knowledge base expands, the mists of uncertainty recede. Now there are many who can see the potential benefits. At that time the cycle can enter the next phase, which is the exploitation phase. This is exploitation in the truest sense: *to turn to practical account; utilization for profit*. Exploitation occurs when a new frontier is turned into a practical, solid place to conduct day-to-day commerce. Now it is time for the people to participate and benefit. Opportunity seems to occur at every turn. Benefits materialize from sources not even imagined by the original explorer. The new frontier no longer exists and in its place is an established part of a mature but growing society. It is accepted as if it had always existed. Today hasn't air travel become the most common way to travel long distances?

The length of the cycle varies, but not as much as one might think. In modern times—let us say the last 400 years—the cycle has varied from a minimum of 30 years to a typical length of 50 years from the initial discovery to the start of the exploitation period. This necessary time period has not accelerated at the same rate as technology. The reason for this is not a function of technology capability but rather a function of human characteristics. These characteristics have not changed dramatically throughout the centuries even though the state of accumulated human knowledge has changed. The proportion of dreamers and visionaries to those who are dubious of new ideas remains the same. As a result, the course of progress is strewn with the obstacles of doubt. As the level of technological frontiers continues to rise, the decision-making time remains fairly constant. The length of time required to reach the exploitation phase of the cycle is really dependent upon how clearly and convincingly the visionaries can paint a picture of reward to those who must make the developmental investments.

The New Frontier

Through the years the peoples of the earth have explored every continent, walked the beaches of every island, climbed the highest mountains, and peered into the depths of the sea. The last great frontier for mankind to explore is the heavens above. So where is our new frontier—*space*—in this cycle? The term "space" may be somewhat misleading because even though it appears to be a great void, it contains many things: the earth, our moon, the other planets, asteroids, the sun, and billions and billions of stars in the heavens. This frontier is so enormous we can't even conceive of its vast limits. Let us look at how our generation is reaching across the void to exploit this grand frontier.

The exploration of space began on October 4, 1957, with the launching of Sputnik, and reached its pinnacle on July 20, 1969, with the first manned landing on the moon. That is the recent history most of us remember. It was an extremely difficult achievement that earned the praise and respect of the world. Even though there is continuing exploration, from that moment on, space was in the development phase. In the mind of the public, however, the goal was accomplished and then the questions began. What is the moon good for? What will space do for me? What are we going to get for our investment? Were the moon rocks worth all those billions of dollars?

I knew that the romance of exploration in space was over during a moment of truth in Bastrop, Louisiana, in the fall of 1969. I had given an after-dinner talk to the Lyons Club on the Saturn/Apollo lunar landing and was answering questions when an old farmer stood up in the back of the room. His question went like this: "Now, young man, I think its fine that we sent men to the moon, but what I want to know is when are you fellows going to figure out how to make a good septic tank?"

Over the ensuing years, exploratory space programs have been very difficult to sell to Congress. Developmental programs have not fared much better. Compromise and funding restric-tions for the Space Shuttle forced the selection of a hybrid, partially reusable configuration, that has experienced delays, a tragic accident, and high operational costs.

Funding delays and the resulting cost escalation have caused serious delays in the development and operation of the Space Station, which is so badly needed for our country to gain the experience and knowledge necessary to be able to live and work in space. What happened is what normally occurs during the development phase. The excitement is gone and there is no clear understanding of how additional investment will lead to practical benefits. Only a few dedicated and patient visionaries keep things going by looking beyond to see the benefits of more knowledge and experience. Knowledge and experience are the pathways to commercial development. A profitable commercial venture, to succeed, must be reduced to routine operations; it cannot be a high-risk adventure. The necessary experience and knowledge in space is being slowly acquired.

In some areas of our space endeavor the exploitation phase is underway, putting space to practical use. Most of us watch the Olympic games live from anywhere in the world. The fall of Communism was brought into our living rooms as it happened; the breach of the Berlin Wall occurred before our eyes. War in Kuwait was carried onto our TV screens as smart bombs found their way down ventilation shafts and Scud missiles were shot out of the sky. The instantaneous spread of news and communication around the world is primarily due to the ability of communication satellites to literally blanket the earth with their coverage. If you place an international phone call, it most likely will go via satellite. The deployment of US satellites is a private enterprise business under governmental control with the cost of space launch services paid for by the satellite owners. It is now a big and profitable industry and getting bigger. From this first big commercial exploitation of space all of us reap benefits every day.

In the public service area, the Landsat program provides so much information on our land and crops that we cannot analyze all of the data. Weather satellites give us a continuous view of the goings on in our atmosphere. The military depends on space for earth observation, weather, communications, navigation, and who knows what else.

The achievement in 1991 of worldwide, 24-hour-a-day coverage of the Global Positioning System (GPS) has revolutionized navigation for those who travel the oceans of the world. During the war with Iraq many of these units were rushed into production for the troops serving in Desert Storm. With GPS our troops knew precisely where they were at all times as they moved across the limitless, featureless desert.

The early investment in exploration and development of our new frontier is now starting to return solid dividends. This is just the beginning. The potential of space is as large as space itself.

Let us close our eyes to the mundane moment-to-moment problems that continuously surround us and let our minds wander freely into this new high frontier. Where is it going? What can we do there? How can we use it to enhance our lives? How can we use it to benefit us as individuals, businesses, and nations? Can we use this new frontier to solve some of our current problems? The time is ripe; the cycle has run its course and is ready for massive utilization for practical account. What visions do we see unfolding before us?

(End of the excerpt from **SUN POWER**)

• • • • • • • • • • • •

We are in the development part of the cycle for space. There is much more exploration to be done, but the great excitement of the earlier period can never be felt again and those that try to recoup the enthusiasm are doomed to fail. What is needed now is to develop the commercial uses of space that leads to the exploitation of space and the great returns that can follow. That is why the suggestion of a new Department of Space to take on the real opening of space for our long term benefits is so important. There is a role for NASA in continuing exploration, but not the development of the infrastructure that is so essential for the commerce of space. It is time for us to realize what must be done now and into the future and go forward.

11   What is Next?

The first and biggest stumbling block to developing solar power satellites is low-cost space transportation. As I described in Chapter 6 this would not have been such a big hurdle if the fly-back booster based on the S-IC first stage of the Saturn V moon rocket had been selected as the Space Shuttle configuration. However, it wasn't so we now need to look at the options that are available today. Fortunately there are several.

One is Space Island Group's approach which was described in Chapter 3. Their approach would short circuit many of the steps described in this chapter and would move directly to commercial development of the system to supply the first electric energy from space. By using elements of the Space Shuttle they would pay for the launch costs of launching the first satellite hardware to low earth orbit by leasing out space in commercial space stations composed of modified external tanks taken on to orbit. They would then assemble the satellites in low earth orbit and move them to geosynchronous orbit using the electrical energy from the satellite to power electric ion thrusters. Their injection of the External Tanks into orbit as commercial space stations would minimize the initial development costs for launching solar power satellite hardware and stimulate the development of reusable launch systems that would be required

as the space station market became saturated.

● ● ● ● ● ● ● ● ● ● ● ●

How would you like to be able to take an elevator to space? Wouldn't it be exciting to be able to step through a door and press the up button for geosynchronous orbit and feel the surge as the elevator car started the upward voyage. Imagine being able to look out the window as the elevator whisked through the clouds, bursting into bright sunlight and then watch as the sky turned dark and then black as you pass out of the atmosphere seeing the stars pop into view. See the panorama of one of the earth's great oceans spread out below. Off to the horizon the earth curves away with an arc of light as the atmosphere becomes a thin band. As you passed through the area of low earth orbit you might even have seen some of the low earth orbit satellites whizzing by or perhaps the International Space Station in all its majesty. The trip would not be short. The elevator would have sleeping quarters, a galley and all the other necessary amenities for a two week trip to geosynchronous orbit as the elevator climbed a ribbon through the black emptiness of space.

When you arrive at geosynchronous orbit the elevator would enter into a giant facility. There would you would find hotel rooms, lounges, and restaurants. Also warehouses, storage facilities and assembly areas for solar power satellites. If you were the adventurous type you could switch to a space ship that is attached to another elevator and go on up the elevator ribbon. When you reach the right altitude and wait for the right time to release the space ship from the elevator ribbon, you would be flipped through space on your way to the moon.

If you were really the adventurous type you may have taken another space ship on out to the end of the ribbon, 62,000 miles above the earth. When you let the space ship go you would be on your way to Mars. You think this sounds crazy? maybe not.

Many years ago a Russian mathematician first proposed the idea of a space elevator, but it was not practical at the time because of the lack of a strong enough material. Arthur C. Clarke popularized the idea in a science fiction novel and then the Japanese discovered carbon nanotubes. Now suddenly there was a material that was potentially strong enough to make the concept work.

Bradley C. Edwards and Philip Ragan in their book, **Leaving the Planet by Space Elevator**, lay out the plans for developing an elevator to space and the benefits it can provide. Bradley Edwards is one of the leaders working on the system. The concept is to deploy a carbon nanotube ribbon that stretches from the earth out into space for 62,000 miles with a counter weight on the end. This would rotate with the earth and maintain itself stretching into space. The end of the ribbon on the earth would be attached to a mobile water based facility so the ribbon could be moved to avoid being hit by low earth orbit satellites. The ribbon would be composed of carbon nanotubes with a strength of 140 to 180 times that of steel for the same mass. As a result the ribbon would be about as thick as a sheet of paper and about 2 meters wide. The carbon nanotubes by themselves have a strength that is 400 times that of steel for the same mass. The challenge is to weave or spin or consolidate the nanotubes into a ribbon of the required strength. This has not yet been accomplished, but they expect it will be done in two to five years.

With the ribbon in place cargo or people can be carried into space by elevator cars that are really climbers. They would have rollers similar to an old fashioned wringer washing machine that will ascend the ribbon powered by electric motors driven by energy beamed from earth with lasers or another wireless transmission system. The cost of transportation to space will be dramatically reduced. There would be a major facility at geosynchronous orbit where solar power satellite hardware could be unloaded and easily moved to the assigned orbit slot.

When we have the Space Elevator there will be no question about the economics of solar power satellites. The question remains can this concept move from a science fiction dream to real hardware?

• • • • • • • • • • • •

Another approach follows a much more conventional path. It is the development of a new fully reusable rocket powered space transportation system. This idea harks back to NASA's original goal for the Space Shuttle. The difference now is we understand the problem much better today, and the technology has improved significantly. The keys to making a space transportation system low-cost are typical of economical transportation systems. First of all they have to be fully reusable. An airline wouldn't be in business very long if they bought a new airliner to fly passengers from New York to Hong Kong, threw it away and bought another new one to make the trip back. But that is exactly what is happening with space transportation today. Another important feature is to have a reasonable sized payload so that the cost is spread over many uses. This allows its capital cost to be amortized over many trips. The cost of maintenance and operations must also be low. Last but not least the cost of fuel must be minimized. The reason these factors have not already been done for space transportation is that there has not been a sufficiently large market to justify the extra development costs involved. The transportation market to launch solar power satellites certainly is big enough for this to happen. In addition the system needs to be designed and developed by a commercial company that

understands all of the costs that must be controlled for it to be a successful commercial venture. If it is done by a government bureaucracy like NASA it probably won't succeed. We have the Space Shuttle as an example of how far off base from a commercially viable system NASA has strayed.

Let's imagine we were given the task of defining a new launch system. The first question to answer is; To what orbit are we going deliver the payloads? The satellite is going to be operating in geosynchronous orbit, which is 22,300 miles above the earth. There are three basic options to get the material for the satellite to geosynchronous orbit. The first is to bring the material to low earth orbit and assemble the satellite there. For reference, the International Space Station is in low earth orbit at about 250 mile altitude. If the satellite is assembled in low earth orbit it could then be moved to geosynchronous orbit using its own solar array to power ion thrusters that would raise it to geosynchronous orbit. The second option is to deliver the hardware to low earth orbit to a cargo transfer facility. From there the hardware would be moved to geosynchronous orbit for assembly there. This would require a vehicle that would shuttle back and forth from low earth orbit to geosynchronous orbit. It could use chemical fuels or be an ion thruster vehicle that has its own solar panels for power. The third option is to launch the hardware all the way to geosynchronous orbit with the launch vehicle. This would require at least three stages. The majority of studies have concluded that launching the hardware directly to geosynchronous orbit is the most costly method and it was better to have the basic launch system optimized for launch to low earth orbit, providing for either low earth orbit assembly or for transfer of the payloads to the higher orbit with an orbit transfer system. As a result the space transportation system that we will concentrate on here is one that can launch the satellite hardware to low earth orbit.

First of all we know the system must be reusable, so that means it has to be able to return to the earth in essentially the same condition as when it took off. The tough part of this is the return trip because of reentry heating. We have a couple of choices. The reentry can be with wings similar to the Space Shuttle orbiter or some type of ballistic reentry device similar to the Apollo capsule. Then we must decide how many stages should be used and what fuels should be considered. Another consideration is how large a payload it should carry. Generally speaking the larger the payload the lower the cost per pound but then development cost and acquisition costs are higher.

Let's start with the type of fuel as it can effect the other options. Hydrogen is the highest energy fuel per unit mass of all fuels that could be used and the product of combustion is water. So it is a very desirable fuel, except for a couple of drawbacks. It has very low density even in the liquid form and its primary source at this time is extracting it from fossil fuels with natural gas as the most common source. However, it can also be extracted from water with the use of electricity. One of my high school chemistry experiments was to separate hydrogen and oxygen from water with electrolysis. When we have electricity available from a renewable energy source then it is reasonable to extract hydrogen from water. Hydrocarbon fuels such as liquid methane or RP-1 which is similar to kerosene and used in the first stage of the Saturn moon rocket, have reasonable energy levels per unit of mass and high energy levels per unit of volume. Their drawback is they produce carbon dioxide when they are burned. Solid fuels have low energy levels per unit of mass, but very high levels of energy per unit of volume. Their drawbacks are they produce high levels of pollution when burned and require mixing and casting the propellant, which is a little like mixing and pouring cement, to refuel. They are also fairly high priced on a per pound basis. They are ideal for military missiles because the high energy per unit volume results in small missiles and they can be stored in the fueled condition for long periods of time. There are other fuels that can be used and in some cases have been used in the past, but they all have significant drawbacks of cost and toxicity and are not appropriate for large scale use.

The least practical fuel would be the solids because of high cost and high pollution levels. They would not be suited for large scale, low-cost space transportation, even when recovered and reused like they are in the existing Shuttle. That leaves us with hydrogen or hydrocarbons as the viable options.

Through the years there have been many studies of launch vehicle configuration design. The most comprehensive studies concluded that two stages to reach low earth orbit was the best configuration. Single stage to orbit vehicles, sound like a great idea. But they require high technology and result in large vehicles with a low payload for the amount of fuel required, which has to be hydrogen. NASA's X-33 single stage to orbit was not successful. More than two stages complicates the system and raises its cost. Stage and a half systems work as shown by the Space Shuttle configuration and the original Atlas vehicle used in the Mercury Program, but because one of the stages must burn fuel all the way to orbit the propellant tanks are big. It still looks like two stages is the best approach.

With that decision we can look at which fuel to select. I say fuel because the oxidizer for either hydrogen or a hydrocarbon fuel will be liquid oxygen. It is clear that the upper stage of a two stage system benefits greatly from the high energy content of hydrogen with minimal penalties for the increased volume required. The other side of the coin is the first stage size can be significantly reduced using the higher energy per unit of volume available from a hydrocarbon fuel compared to hydrogen and the cost of the

fuel is less. So at this point in time I would select a hydrocarbon fuel for the first stage. Later on when there is energy flowing from space that can produce hydrogen from water it would be desirable to replace the hydrocarbon first stage with one powered by non-polluting hydrogen even though it would be bigger.

I have been talking about stages, but I'd like to call them something more descriptive. I call the first stage the booster and the second stage the orbiter which brings it more in line with what we are used to from the years with the Space Shuttle.

Next is the question of what type of reentry the system should it use. If both the booster and orbiter use the winged approach, we have a good example of what the orbiter might look like from the existing shuttle orbiter. It would be much larger as it would need to carry its fuel and oxidizer within the body of the vehicle as well as the payload. It does not need any secondary power for landing as the shuttle orbiter has so ably demonstrated. The booster is a different story; it will need some method of returning to the launch site after it has staged the orbiter. The most straightforward method is to provide it with jet engines for the flight back to the launch site after reentry. An alternate approach that has been suggested is to have the booster land down range without power and then have an auxiliary jet engine pack that could be installed for flying back to the launch base. The problem with this approach is the added time required and the complexity of the system. I am going to stick with the flyback jet engine approach. The speed required for staging also affects the booster design. The most desirable is to have it as high-speed as practical and still be able to make it a heat sink booster like the winged version of the S-IC. This would not require any thermal protection system. It should be able to fly back from a booster mission, land and be ready to be mated to another orbiter and refueled for the next launch in less than 24 hours. This scenario assumes a

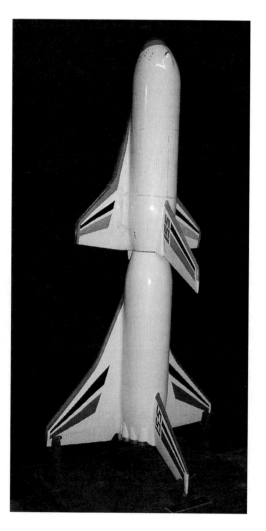

vertical launch for the booster. I will discuss the potential of using a horizontal take off in later paragraphs.

The next question to address is what does the orbiter look like and how should it be mated to the booster? It is interesting to note that starting with a heat sink booster and then developing an orbiter design that carries all propellants internally as well as a large payload bay, the two vehicles will be nearly the same size. They will have similar shapes that follow the configuration developed for the current Space Shuttle orbiter. The booster will have a larger wing and is a heavier vehicle. Because of extreme reentry temperatures the orbiter will have a thermal protection system for orbital reentry. The options for mating the vehicles are belly to belly, - or - the orbiter on the back of the booster, - or - nose to tail. I prefer the nose to tail mating arrangement as it keeps the orbiter completely away from any possible damage from ice or debris damage during launch. This allows the two vehicles to be joined in the horizontal position while the vehicles are on their own landing gear. The nose of the booster would be a dome that rotated 180° to provide

Figure 6: Model of two stage reusable launch vehicle developed during the Solar Power Satellite System Definition Studies. Hydrocarbon fuel Booster with a Hydrogen fueled Orbiter 1 Million pound Payload

room for the orbiter engines when in the mated position and after separation would rotate back to form the booster nose for reentry and the flight back to the launch base. The mated vehicles would be raised to a vertical position on the launch pad.

The following illustration is a model of the configuration that was developed during the System Definition Studies. It was designed to have a million pound payload. The booster used hydrocarbon fuel and the orbiter used hydrogen fuel. A smaller version was also developed that had a two hundred thousand pound payload capability.

The booster was designed to be unmanned, while the orbiter was designed to have a crew and passenger capability.

Another way to make the system smaller for the same payload would be to design the booster to carry its fuel in the wing with only the liquid oxygen in the body of the vehicle. If you were really brave and didn't have the chief engineer of a NASA center to stop you, the liquid oxygen for the orbiter could be carried in the wing.

Another winged concept that has great potential, particularly for smaller payloads or personnel launches would use a horizontal take off configuration. This would be like Burt Rutan's White Knight and SpaceShipOne only much bigger. The booster stage would be a large jet powered airplane with a rock-et engine in the tail carrying a hydrogen powered orbiter on its back or underneath if it was a twin body design. The combination would take off from a runway as a jet airplane and climb to 40,000 to 50,000 thousand feet where it would light off the rocket engine. Then it would accelerate the two vehicles to about Mach 3 (a little over 2,000 miles per hour) at about 100,000 feet altitude. The orbiter would then go on to orbit while the booster airplane would decelerate to sub-sonic speed and fly back to the launch base.

The alternate to winged reentry is a ballistic reentry concept. This can be accomplished with the stages entering in a variety of positions as long as the part that enters first is protected with a thermal protection system and is aerodynamically stable. The landing can be accomplished with parachutes similar to the way the Russian Soyuz or the Apollo spacecraft landed on land or in the water. Or it could have auxiliary landing rockets like the concept developed for the DC-X design proposed by the former McDonnell Douglas Company. This is one of the designs being considered by the Space Island Group for a future system. During the Solar Power Satellite System Definition Studies one of the designs considered, used a booster and orbiter system that were shaped like an elongated Apollo capsule that reentered engines first. The engines were buried in a heat shield that used water converted to stream by the reentry to protect the vehicle from reentry heating. Small rocket engines provided for a controlled water landing. The main issue was protecting the vehicle from water entering the system after landing to limit the amount of refurbishment required before the next flight. The booster would land downrange from the launch site and be towed back. The orbiter on its return from space would land adjacent to the launch site. The illustration is a model of this vehicle. The booster stage uses hydrocarbon fuel and the orbiter uses hydrogen. The payload bay was expanded for the launch and after the payload reached orbit the yellow portion shown in the model telescoped into the upper nose section so that the reentry shape was very much like an Apollo capsule. The nick name we gave this vehicle was "The Big Onion."

All of these reusable systems would cost dramatically less to launch than throw-a-way systems. The other factors that must be incorporated into the system to maintain low operational costs would be to design the vehicles for low maintenance requirements between flights and to provide for rapid loading and unloading of cargo. The use of space flight low-weight standardized cargo containers would be essential to achieve this goal. All of the rocket powered systems would then be designed to go to a space base cargo handling facility in low earth orbit

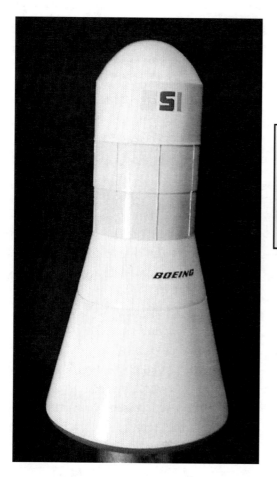

Figure 7: Two stage ballistic recoverable, reusable launch vehicle. Hydrocarbon fueled booster and a hydrogen fueled orbiter stage. Yellow payload bay telescopes down for reentry. Engines are buried in the heat shield in the base of each stage. 1 Million pounds payload.

where the cargo would be transferred to other vehicles designed to move it to geosynchronous orbit where the satellites would be assembled. These orbital transfer vehicles would most likely be powered with solar powered ion engines to keep the costs low.

An alternate to assembly of the satellite in geosynchronous orbit is to assemble them in low earth orbit and move the completed satellite to geosynchronous orbit using ion engines powered by the satellite's solar array. One of the concerns for low earth orbit assembly is the problem of all the space junk out there that could damage the satellite during assembly or during the transfer to the higher orbit. In addition the large area of the satellite as it nears assembly will cause some drag, even in the near vacuum of space that will require some energy to maintain its altitude just as the International Space Station.

One of the steps that needs to be taken is to reach international agreements to ban any launch systems that leave junk in space, such as explosive bolts. There should also be a systematic cleanup of all the current junk that orbits the earth. Currently there are thousands of objects large enough to track that are a danger to future systems in low earth orbit. Some of these objects are at a low enough altitude that their orbit eventually decays and they burn up on reentry. In the meantime many more are in higher orbits and more are being added all the time. It is a serious problem that needs to be addressed.

In the early months of 2009 there was a collision of an Iridium communication satellite and a Russian satellite at about 500 miles altitude. This resulted in a massive field of space debris that will be a danger to orbiting spacecraft far into the future if it is not cleaned up in some way.

• • • • • • • • • • • • •

I have taken the time to describe many of the options available to actually place solar power satellites in space because it is the greatest cost element of developing and building the satellites. It is the main reason they have not yet been built. In order for a commercial aerospace company to invest the enormous cost of developing a reusable launch system they must be assured of a large enough market for their vehicle in order to amortize the development cost, the cost of the vehicles, and to make a profit. To give you an example of what this means I will use the Airbus A380 jumbo jet as and example.

Airbus estimated the development costs of the airplane in the neighborhood of $12 billion dollars. They originally estimated they would have to sell 250 airplanes in order to break even. After they had some development problems that delayed introduction of the plane for a couple of years, that number increased to about 450 airplanes. The cost of developing a heavy-lift reusable space freighter could be as much as twice the cost of developing the A380, so you can imagine the size of the market required to justify commercial development of the system. The only commercial market in the short term with that potential is building solar power satellites. Eventually space tourism and colonization will develop another kind of market, but the need for the solar satellites is here now and they will open the other space markets to mass use.

I haven't talked about one other option, but I think I need to bring it up. That is the option of the government funding the development of a space transportation system that would be a basic piece of the infrastructure needed for solar power satellites. The United States has spent hundreds of billion of dollars on the Iraq War and face a future of more wars over Middle East Oil that can cost us trillions of dollars. The United States Space Security Agency has conducted a study of their own on Space Based Solar Power and one of their conclusions is the United States can save a half a trillion dollars a year in war cost avoidance by developing Space Based Solar Power. It seems like $25 billion to develop a reusable launch system for solar power satellites would be a wise investment in our future. It would have to be done under a new agency other than NASA that hasn't had time to develop into a bureaucracy. The Department of Space suggested by the Aerospace Technology Working Group, would be ideal.

The advantage of Space Island Group's approach is much of the cost will be paid by commercial users for space manufacturing and accelerated space tourism. In addition they will be building most of their system around hardware that already exists. Their concept converts a major part of a normally expendable element of a launch system, the external tank, and turns it

into a useful income producing component in space. The problem with their approach is its success is limited by the number of commercial space stations that can be leased out before the market becomes saturated. However, it has the potential to get the program started and thus spur the development of new space launch systems.

As in all new industrial developments timing is of the essence. Bill Gates and Paul Allen started Microsoft in a garage with a small amount of funding. They rapidly developed one of the largest companies and industries in the world. It is not likely that the solar power satellite industry can be started with such small funding because the initial hurtle is so high, but timing will be critical and the launch system will pace the implementation. Space Island Group has the starting edge if rocket systems are to get the program underway. However, if the Space Elevator development moves as fast as Bradley Edwards predicts, it will eclipse earth launched rockets. If these efforts stumble, the existing aerospace industry can jump in to fill the need. This will leave the door open for innovative new entrepreneurs like Burt Rutan.

The point is there are many options for delivering solar power satellite hardware to space that all have the potential of meeting the required cost goals. Which it will be has not yet been determined. Timing is of the essence as to who will capture the job and create a new industry and be the Bill Gates and Paul Allen of space transportation.

On the day after election day 2006, the news of the day was the sweep of Congress by the Democrats, but even eclipsing that news was the resignation of Donald Rumsfeld as Secretary of Defense. He was forced to resign as the war in Iraq, which he had directed, continue to claim the lives of American service men in a war that did not have an end in sight. The American people had voted for a new direction that would get us out of the war. It brought back vivid memories of the first time I saw Donald Rumsfeld.

It was 1975. I had been back in Seattle from New Orleans for a couple of years following our failed bid to build the external tank for Space Shuttle. During this time I had been assigned as the manager of the Boeing Design to Cost Laboratory. A job I did not want but it was an interesting and educational job until I could get back into a space program. The Saturn/Apollo program was over, we had lost out on the Space Shuttle and the external tank manufacturing job. Our space future as a company was cloudy. But NASA was working on solar power satellite studies and I was trying to get transferred back into Boeing's advanced space programs so I could work on them. That was when I received a call summoning me to the office of the senior Vice President of the Boeing Aerospace Company. When I arrived he sat me down and started telling me what he wanted. He began

with an outline of how the various aerospace companies cooperate with the Defense Department in providing them with experienced personnel. He reminded me that earlier when he was with Boeing he had taken a position in the Department of Defense and just recently returned to Boeing as a Vice President. This kind of shuffling of people from industry to the government and back was a common practice.

He went on to tell me that the Air Force was looking for someone to fill a position for long range planning at the Pentagon and he wanted me to apply for the job. It was for an indeterminate period and he could not make any guarantees, but the chances of me returning to a high level management job at Boeing was very good. He slid a piece of paper across the desk and said, "Here is the guy you are to meet with for an interview." The paper also specified a date, time and location. There wasn't any option offered for me to turn the proposition down.

I went home that night with very mixed feeling. I didn't want to work for the government, but it was also a great career opportunity at a time when there wasn't much else on the horizon. My wife and I debated what to do and in the end I was on an airplane to Washington DC.

The morning I was scheduled to have the interview I caught a taxi from my hotel to the Pentagon. As the cab approached the Pentagon the driver was mumbling about the crowd of people milling around outside. He finally stopped and told me he couldn't get any closer and I would have to walk the rest of the way. I was working my way through the crowd towards the entrance when a voice came booming over loud speakers scattered around the area. It was then I realized there was a podium set up near the entrance. The speaker was trying to get the crowd's attention. When they finally settled down he started in with an introduction that ended with, "and here is your new boss Secretary of Defense, Donald Rumsfeld." By this time there I was with briefcase in hand standing very near the podium as Rumsfeld came to the mikes and started his speech. He had just been appointed by President Ford and was the youngest man ever to hold the position. I stopped to listen as he spoke to the assembled crowd that had left their offices to come hear his talk. It was about time for my interview, but it was obvious that this unexpected event preempted any normal schedule so I stayed and listened to the end. I didn't much like what I was hearing. His words added to the internal struggle I was still going through about what I was going to do about the interview I was about to have.

When he was through I made my way in through security to the appointed office. Fortunately, it was not my first visit, as I had briefed Air Force personnel on Design to Cost several times over the past two years so I knew

my way around fairly well. It is a huge place and easy to get lost. I actually arrived at my appointment before my contact could return.

We talked for over an hour and he described the job they were offering. It sounded interesting, but it just didn't strike a chord within me. The comments I had just heard from Rumsfeld added to my unease. I kept thinking, if I take this job I will lose any chance to develop solar power satellites and will be in an organization led by an unknown individual that had left me with an uneasy feeling. By the time it was over I had made up my mind. I turned him down. It was a feeling of relief as I took a cab from the Pentagon directly to the airport and caught an early flight to Seattle. In those days you could change flights right up until the gate was closed and I was an expert at changing flights because of the endless flying I had done on the Saturn and Shuttle definition programs.

When I reported back in Seattle that I had turned the job down they were not happy. I let things be for a week or two and then went to the President of Boeing Aerospace Company and reminded him he had promised to let me return to the space program after I had established the Design to Cost Lab. I also reminded him that they had been ready to let me go to the Department of Defense. He had been happy with our Design to Cost results and used our data extensively to make points with the Air Force so he said "OK, you can go back to space." I took over as Manager of the Boeing Solar Power Satellite Program.

• • • • • • • • • • • • •

Today as I look back at the years of war we have experienced in the Middle East and realize they could have been avoided if we were not dependent on their oil, it makes me sad. The turmoil in Iran goes back to the days before the Shah, when U.S. oil companies controlled their oil production. The situation started to unravel in 1951 when Mohammad Mossadeq gained control, became the Prime Minister and nationalized Iran's oil. The United States intervened in 1953 when the CIA supported a military coup to returned the Shah to Power in order to attain a more favorable oil policy. This was followed by the Arabian American Oil Company (ARAMCO) agreeing to return 250,000 square miles of oil concessions back to Saudi Arabia in 1963 and the nationalization of Iraq oil in 1972. The oil embargo of 1973-74 was triggered by the United States and other Western nations support of Israel. So oil was being used as a weapon of war against the West. There were many people and some politicians in the United States that were demanding we send in the Army and take the oil. Didn't we have a right to it?

• • • • • • • • • • • • •

During the period that I was working with congress to pass the Solar Power Satellite development bill in 1978 and 1979 I had many sessions with Senator Henry (Scoop) Jackson from the state of Washington. He was head of the Senate Energy Committee. One of his concerns that he spoke of often was the Straits of Hormuz. This narrow access to the Persian Gulf between the United Arab Emirates and Iran is the primary route of Middle East oil shipments. His worry was that at some point some country or terrorist group would realize how easy it would be to block off access to the Persian Gulf by blocking the Straits of Hormuz. If this happened we would be in real trouble. He recognized how much our economy and standard of living depended on Middle East oil.

To try and prevent this from happening the United States maintains a significant military presence in the area with ships and aircraft. So even in peace time we were spending money for the military to maintain the flow of oil. None of this ever gets counted in the cost of oil.

During World War II bombing raids on the Ploesti oil fields in Rumania were primary targets for the Allied bomber fleets as they tried to stop the flow of oil to the German army and air force. The loss of their oil supply caused a great disruption for the German forces and Germany had to resort to extracting fuel from coal. In the United States fuel rationing was necessary in the civilian population in order to conserve oil needed for the war effort.

• • • • • • • • • • • • •

In 1990 Iraq invaded Kuwait to gain control of their oil. The United States led a multi-nation coalition in the first Gulf War to free Kuwait and reclaim their oil. As the Iraqis were driven out of Kuwait their parting move was to set fire to over 500 of Kuwait's oil wells. The news was full of pictures of the giant flaring plumes of fire and dense black smoke lining the horizon. The view from space was even more startling as the smoke covered much of Kuwait and the Persian Gulf until the fires were finally extinguished.

After freeing Kuwait the United States military maintained a strong visible presence in the Middle East. This presence fueled the terrorists' resentment against the United States and led to the attack on the destroyer **Cole** in Yemen with the death of several U.S. sailors. On a much larger scale it was probably a major cause that led to the 9/11 attack on the United States in 2001 that has had such a devastating impact on the lives of every American and also the rest of the world. Our reaction was the invasion of Afghanistan as we pursued Osama bin Laden and his Al Qaeda supporters. Osama bin Laden is still free and the news today is about renewed fighting in Afghanistan and the threat of new attacks on the U.S.

Our invasion of Iraq was advertised as a war against terrorists and a dictator with weapons of mass destruction, but we soon found out there were no weapons of mass destruction. We also found no tie between Iraq and the Al Qaeda terrorists. What it was really all about was protecting our access to Middle East oil. The Bush administration was afraid that if we backed away from Iraq and left them to their own devices the whole region would topple into instability that would cut off our access to their oil.

Now we are facing a situation in Iran that is extremely serious. Energy is involved here also as Iran is working to develop nuclear power. The difficulty is they are suspected of developing nuclear weapons as well. The Bush administration also accused them of helping the insurgents in Iraq. I suspect that they are interested in developing nuclear power because they see the end of their oil. If this is the case one can understand why they need to develop a new source of energy. We must be very careful how this situation is handled or we will find ourselves in yet another war in the Middle East.

The change in power in congress in the 2006 election and the forced resignation of Donald Rumsfeld as Secretary of Defense changed the direction of the war in Iraq, but the real solution is to eliminate our reliance on Middle East oil. We have to replace oil as our primary energy source.

We obviously cannot do that overnight, but if we were to show the world that we were solidly set on the course of replacing oil there would be a dramatic shift in the way the oil producers looked at their future markets. If they looked into a future that did not need their product they would do whatever they could to make it more attractive and diffuse the competition. The economies of nearly all of the Middle East nations are totally dependent on oil. Oil and natural gas are their only exportable products. Oil money is pouring into Dubai in the United Arab Emirates. They realize they only have ten years of oil left and they have built up an impressive financial, business, and tourist industry. They are already planning on a future without oil.

If the world continues to rely on oil for energy we will be facing more wars either in the Middle East or more wars over who will have access to Middle East oil. After we pass the peak of world oil production there will be increasing competition for the oil that remains. This will be a monetary competition to start with, but it is extremely likely as the costs become higher and higher some country will become desperate enough for energy to use force to grab it, just as the Japanese did on December 7, 1941.

We can avoid these kinds of wars if we are willing to make the investment now to develop solar power satellites. This is the only energy system that has the potential capacity that can be delivered to any nation on earthl. We have already spent nearly a trillion dollars on the war in Iraq. If that had been spent on developing and building solar power satellites we would be well on our way to the fourth energy era right now. If we consider the development of an energy system to replace oil as the

cost of averting war over oil, the price would be really cheap. How much are you willing to pay for the lives of American soldiers and innocent civilians? We have a chance to make the choice by demanding that our government act now to lead our industries forward into developing solar energy from space.

On October 10, 2007 The National Security Space Office released their interim report on a study of Space-Based Solar Power. This report was based on a study that they had made through a very innovative process that tapped the talents of 170 experts through a web page on the internet. The report can be summarized by their four overarching findings/themes. (The underlining and bold print was part of the summary) They are:

1) Space-Based Solar Power **does present a strategic opportunity** that could significantly advance US and partner security, capability, and freedom of action, and **merits significant further attention on the part of the United States Government and the private sector.**

2) While significant technical challenges remain, Space-Based Solar Power **is more technically executable than ever before** and current technological vectors promise to further improve its viability. A **government-led proof-of-concept demonstration** could serve to catalyze commercial sector development.

3) SBSP **requires a coordinated national program with high-level leadership and resourcing** commensurate with its promise, but at least on the level of fusion research or International Space Station construction and operations.

4) Should the U.S. begin a coordinated national program to develop SBSP, it should expect to find that **broad interest in SBSP exists outside the US Government:**
* **Aerospace and energy industries**
* **Japan, the EU, Canada, India, China, Russia, and others**
* **Many individual citizens** who are increasingly concerned about the preservation of energy security and environmental quality.

5) While the best chances for development are likely to occur with US Government support, **it is entirely possible that SBSP development may be independently pursued elsewhere without U.S. leadership.**

The interim report is approximately 85 pages long and explores the potential of Space-Based Solar Power to provide energy to remote battlefield sites that could significantly reduce the cost of providing needed energy and at the same time saving many lives. It considers the capability of providing emergency power to disaster areas. It expands on the ability of solar power satellites to reduce our reliance on Middle East oil and thus dramatically reduce the threat of future wars. That alone could save this country a half a trillion dollars a year and uncounted lives.

This study is quite amazing because it was initiated and conducted by military personnel with no cost to the government. It illustrates the potential impact space-based solar power will have on the people of the world once they understand its potential and the impact it will have on their lives in the future.

By developing solar power satellites we not only eliminate one of the great reasons for war for our country, but also for other countries. Energy from the sun is not a resource that can be depleted for as long as mankind is alive. It can be made available to all the nations on earth. We do not have to fight over a resource that has limited capacity. We can use technology to solve many of our problems rather than technology to kill one another. Also, in developing solar power satellites we will be developing the capability to have ready, low-cost access to space and all of the other potential resources it contains. We would open up the potential to mine the asteroids and establish colonies on the moon and explore Mars and the moons of Saturn or Jupiter. The possibilities are endless with low-cost access to space.

Today as James Woolsey, former CIA director stated, "We borrow a billion dollars every working day to import oil, an increasing share of it coming from the Middle East" How much longer can we afford to wait?

13 Bring Energy to the Entire World

Many people of the world live in poverty. There is no hope for them if they cannot raise their standard of living by increased productivity through the use of energy. China and India are building coal fired generating plants as fast as they can to try and satisfy demand. There are now electric outages simply because there is not enough electricity to go around. The addition of so many coal plants is magnifying atmospheric pollution and accelerating global warming.

The demand for electrical energy is growing rapidly so the world market for electricity is reaching toward two trillion dollars a year. This does not count the energy needed to replace oil used in the transportation industries. So this huge market is both a blessing and a great obstacle to overcome. It is a blessing because of the enormous economic potential that will provide the income stream to pay for new generating capacity. The great obstacle is to first develop the system in the middle of a world recession and then to build the generating capacity fast enough to satisfy the growing electricity demand, replace atmospheric polluting systems that are causing global warming, and build up capability to replace oil.

To help you understand the sources of greenhouse gases we find that power generation is 24%, with other energy related systems 5%, transportation is 14%, building heating is 8%, industry accounts for 14%, with the remainder being non-energy emissions with land use 18%, agriculture 14%, and 3% from waste.

Solar power satellites is the only system that has the kind of capacity that can ultimately achieve the goals of stopping global warming and replacing oil as our primary energy source. In this chapter I will describe how they can be developed and deployed to provide energy to the entire world. In Chapter 9 I outlined a development and deployment plan based on an industry/government partnership. It was based on the concept of an orderly development of the system that started with government investment in a ground based demonstrator and government leadership. It left open the question of how much of the space infrastructure development would be government funded versus commercial development. It did specify the need of government incentives like loan guarantees and tax incentives. But much time has past and the need for rapid action is becoming critical, the United States needs to make the commitment to proceed if we want to avoid disaster.

The most crucial step in parallel with a demonstrator is the development of a fully reusable space transportation system. It is probably too late and the economy is in too bad a shape to expect its development without the major assistance of the government. This may mean that our government fund the development of the system or work in partnership with other nations. It would be better however, if an incentive system could be worked out so that industry does the development. For instance, the land incentives given to the railroad companies to build the transcontinental railroad in the mid 1800s could be a model for today's energy challenge.

If the Space Elevator is successfully developed it will likely take over much of the space transportation function. This is because of its potentially lower-cost and ability to deliver payload directly to geosynchronous orbit without the need to use a second system to move the payload from low earth orbit to geosynchronous orbit. In order to be a successful transporter of solar power satellite hardware it will require multiple Space Elevator ribbons in order to have enough capacity. For each ribbon going up there will need to be some elevator ribbons dedicated to return the climbers from space to the earth. This will be necessary because of the time it will take to make the trip to geosynchronous orbit. It will be like a one way highway going up with multiple cars spaced at appropriate distances apart and then a one way ribbon coming down for the return trip.

The chance of the Space Elevator being developed in time is fairly remote. We need to proceed on the assumption that a rocket powered launch system is the one that will be used to launch solar power satellite hardware.

If the satellite hardware is launched with a rocket powered transportation system the location of the launch base becomes a key element. The majority of United States vehicles have been launched out of Cape Canaveral, with the Air Force launching many of their flights out of California. Cape Canaveral is located at about 28 degrees North Latitude. As a result there is an azimuth correction penalty for satellites being launched into an equatorial orbit. This is the reason that Sea Launch moves their water born launch pad to the equator in the Pacific Ocean for launch. Another problem with Cape Canaveral is the weather and storm delays, which has caused several launch delays because of weather. Another drawback is noise. There will likely be many launches a day with the same number of returning boosters and orbiters, so the noise will be a real problem if the launch site is near a populated area. Think of today's busy airports with really noisy vehicles. The advantage of Cape Canaveral is it is in the United States. Space Island Group is planning on launching from a new launch facility at Cape Canaveral because of the existing infrastructure for their system derived from Space Shuttle elements.

The most desirable launch location for solar power satellite hardware is near the equator where there is a minimum of lightning storms, hurricanes, and other bad weather. In addition the launch windows for reaching a low earth orbit cargo handling facility occurs every one and a half hours, which makes multiple launches a day practical. There are several candidate locations near the equator, but none of them in the United States.

My favorite would be Canton Island in the Phoenix Group of Kiribati. It is located about 2 ½ degrees below the equator and about 700 miles north of American Samoa. While we were cruising in the South Pacific we spent a hurricane season at Canton as hurricanes do not occur on the equator. It is a low lying atoll with the only inhabitants a few native families that are there as caretakers. The island has quite a history. It was set up as a refueling base for Pan American Clippers before World War II. During the war it was fortified by the United States with 30,000 troops. One runway was build as a bomber base and another for fighters. By the time it was completed the war had passed it by and it only suffered one bombing attack by the Japanese. After the war, possession was shared by the Americans in the north and the British in the south. The lagoon had served as a base for Navy PBY flying boats as evidenced by one of them still there, wrecked on the beach. Part of the lagoon had been dredged for ships. After the war Canton was used again as a refueling base for transpacific airline service for DC-6s and Stratocruisers. The planes would make an overnight stop with the passengers staying on board a former troop ship that had run aground in the pass that led to the lagoon while being pursued by a Japanese submarine. The ship was permanently lodged on the bottom and was converted to a hotel after the war. At some point it caught fire and burned. But by that time new longer range airliners were in service that did not need to stop and refuel.

The next use of the atoll was as a base for NASA during the early manned space program. It was from Canton that John Glenn's Mercury capsule was tracked when they thought the heat shield had come loose. Then the United States Air Force took over and converted it to a secret base for tracking ballistic missile testing in the Pacific. All our charts warned shipping not to approach closer than 50 miles. They created a base for their personnel with nice houses, power plants, water desalination facility, telephone system, library, radio station, chapel, machine shops, fuel storage tanks, and recreation facilities. A nice place to be stationed.

When the nation of Kiribati was formed from the Gilbert Islands, the Phoenix Group and the Line Islands in 1979 the United States turned over Canton to Kiribati. It was left unoccupied for many years and it was badly vandalized by visiting fishermen and seafarers. Finally the Kiribati government sent out a small group of caretakers. When we were there in 1991 for four months, there were seven native families living on the island. Canton is the only island in the Phoenix Group that is occupied. The residents received supplies four times a year from Tarawa. It was a memorable experience for us.

We were anchored in the lagoon along with two other cruising sailboats about a mile from the village. The native families lived in the houses built by the Air Force, but there lifestyle was nothing like ours. The pigs lived in the kitchens and they cooked outside in the yards. Sleeping mats were set up on platforms outside that they made from the raised floors that had formerly been in the computer rooms when the Air Force was there. None of the power plants were operating, even though it looked like if you pushed the starter buttons on the huge diesel generators that they would start. The water desalination plant had been stripped of many pipes and fittings. The only water available was one brackish well and rain catchments from the roofs. The problem with that was in the four months we were there we only had two brief showers. The majority of their drinking liquid came from coconuts. Their main food was the fish they caught with throw nets in the lagoon, some scrawny chickens, a few pigs, coconuts, a few papayas, and lobsters. This was supplemented by supplies from Tarawa, but the problem was they were not very good at planning so that supply would be gone soon after it arrived. We all had stocked our boats with basic food for at least six months so we were able to supplement their meager supplies while we were there.

Most of the cruisers carried folding bicycles on board so we could ride to the village or explore the atoll. It was fascinating. There were old gun emplacements left over from the war. There were bunkers hidden under brush that hadn't been entered for 45 years. I retrieved a transformer that would convert voltage from 110 to 220 or 440 or the other way around. It turned out to be very useful during later haul outs to convert the 240 volts that was the common voltage in the South Pacific islands nations so I could run my 110 volt power tools. It had a name stamp that said it belonged to the U.S. Signal Corp with a date of 1939.

The fighter base was overgrown with brush and was lined with revetments to protect parked planes from air attacks. The bomber runway was in excellent shape as it had been used extensively by the Air Force during their use of Canton. The runway is about 7,500 feet long with room so it could be extended. There were pieces of old airplanes that stuck out of the ground where they had been bulldozed at some point in the past. As I rode my bicycle down the runway the image of returning orbiters always flashed through my mind. It would make a great launch base.

The weather never changed with the temperature varying about 6 or 7 degrees from night to day. The breeze was steady at bout 10 to 12 knots. The sun was hardly ever obscured by the few clouds that drifted by. The lagoon was dredged to about 35 feet depth and had a small loading dock and many huge diesel storage tanks. The nearest inhabited island was about 600 miles away. The island is about 7 miles long with plenty of space for housing remote from the launch site, and the fishing is terrific.

Since we were there the Kiribati government has upgraded the airport to allow it to be used for refueling their 737 jet airliners. It has all the characteristics to make it an ideal launch site.

There are other islands near the equator that are also potential launch sites. Nauru, an island nation in the Micronesian South Pacific, is very close to the equator and is a good potential candidate. There is also the French launch base in French Guyana which is near the equator. It is likely that more than one will be developed eventually, but Canton would be a good place to start. The only problem is we need to get started soon, before global warming melts the glacier ice and sea levels raise to submerge these low lying atolls.

• • • • • • • • • • • • •

As development starts on the launch system for solar power satellites it will be essential to start the preparation of the launch base. It will have to have the capability of launching several times a day with multiple launch pads. Let me describe a launch system for a two stage vehicle composed of a fly-back heat sink booster using hydrocarbon fuel and liquid oxygen as propellants. The orbiter will be a hydrogen fueled vehicle with wings that looks a lot like the existing Space Shuttle orbiter except bigger and it carries all its propellant internally.

The payload would be loaded in the orbiter stage in preloaded lightweight cargo containers in the horizontal position with the two mated stages. After being raised to the vertical the stages would be fueled and ready for launch. Built into the stages would be systems that would check out all systems so that by the time the fuel was aboard they would be ready to launch. By being on the equator their launch window would occur about every hour and a half in order to be able to rendezvous with the cargo transfer facility in low earth orbit, also on the equator. After launch, the booster stage would separate and re-enter the atmosphere down range of the launch base. When it had slowed to subsonic speeds it would start its jet engines and fly back to the launch site for a conventional horizontal landing on the runway. It would be allowed a few hours to cool from the reentry heating and then be ready to be mated to another orbiter. All systems would have been monitored during launch and fly-back so that it should be ready to go within 12 hours of its last launch. The pacing time line would probably be the time it takes to fuel both it and the orbiter stage. I would expect a booster could make at least two launches a day. The orbiter would be less because of the time spent in orbit to unload and to cool after reentry and landing. Because of the much higher reentry heating and the need for a separate thermal protection system the orbiter would also need to have a visual inspection after each flight.

The engines for both stages would

be designed for reuse and should not require any significant maintenance between flights. Periodic maintenance would be required, but nothing like the present Space Shuttle. Today's airliners fly for years without removing an engine. Because of the lower flight rate for orbiters it would be necessary to have more of them than boosters. One of the questions that might be asked is, "Will these vehicles have manned crews?" They could be either manned or unmanned. With the state of technology today it would not be necessary to have crews, but it may be desirable. The option is there to go either way.

When an orbiter gets to low earth orbit it will deliver its payload to a payload transfer base. This will be a relatively short stop as all the cargo will be in standard containers. As soon as they are transferred to the transfer base the orbiter then returns to the launch base.

The next step in the transportation sequence is to move the cargo from the low earth orbit base to geosynchronous orbit. This could be done with rocket powered vehicles that shuttled back and forth between the two orbits, but the most likely process would be to use ion thrusters that are powered with solar energy.

When the satellite hardware reaches geosynchronous orbit, assembly can begin. Even though the satellite will be immense it is simple in design. It would be composed of a structural frame covered with vast blankets of solar cells, including the wiring to gather the electricity. The structural frame would be designed to be assembled in space robotically. Huge rolls of solar blankets would be brought to orbit with their wiring harness attached allowing them to be unrolled onto the structural bays. This is repeated over and over again as every bay is exactly like the previous one until the entire satellite array is assembled. The satellite would then be ready for assembly of the transmitter.

The likely first step would be to install the rotary joint that attaches the transmitter to the satellite's solar array. This could be a preassembled module that would be brought to orbit in one piece or possibly in several pieces that could be easily joined into the final assembly. With this component in place on the satellite it would be ready for the transmitter which would be assembled robotically. After the supporting structure is assembled the transmitter sub-arrays would be added. This is a very simplified description of a fairly complex process but very similar to what is used in automobile assembly today. But for our purposes I think you can see what a relatively straightforward process is involved in placing solar power satellites in place.

After the satellite is assembled, there is one other function that must be provided to make the system work. This is the attitude and station keeping system. This system keeps the satellite oriented so it always faces the sun and also stays in its precise orbit slot. The energy to provide this capability would be supplied by small ion thrusters

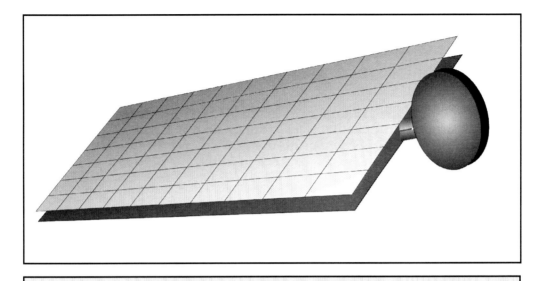

Figure 8 **Assembled Satellite Configuration from Systems Definition Studies**

located at several locations on the satellite and powered by the solar array. The working fluid used by these thrusters would probably be argon or some other inert element that would need to be replenished periodically during the life of the satellite. Speaking of the life of the satellite, it is important to note that they should be designed and built to have an indefinite life. Their life expectancy would not be only 30 or 40 years, but with proper maintenance they would be power plants that will still be providing energy to the earth a hundred or two hundred years in the future. In this respect they are like hydroelectric dams, except they are solar dams.

As I look back to when growing up I watched the construction of Grand Coulee Dam over the years, I can first remember the pouring of concrete for its foundation behind the coffer dams

on the bed rock of the Columbia River in 1935 Then through our entry into World War II in 1941. When I realize the effort and man hours that went into its construction, I am awed to think each satellite will provide the same amount of energy as Grand Coulee Dam.

The assembly process I have described assumes that the satellite would be assembled in geosynchronous orbit. Another option is to assemble the satellite in low earth orbit and use its solar energy for the ion thrusters that would raise the satellite as a complete assembly into geosynchronous orbit. This is the approach the Space Island Group is planning to use.

There are other configurations suggested that will undoubtedly be considered, but they all add significant complications for assembly, operations, or cost. I expect the configuration that will be built is some variation of the reference design. In a commercial development the engineering process will select the simplest and most cost effective configuration.

The development of the first satellite and the infrastructure will likely take seven to ten years. But after that the launch and assembly time for a single satellite can be reduced to about one year as most of the assembly will be done by automated robots. A year's construction cycle would require a launch rate of two or three launches a day with a transportation system that can carry 100 tons of cargo per launch. There could actually be more than one satellite under construction at a time. The controlling factor first of all will be demand for the clean energy and then the pacing control will be manu-

facturing capacity. We only need to look to the current airline industry as an example. Boeing was selling their new 787 Dreamliner at a rapid pace until the economic recession hit. The number of orders was over 900 in the fall of 2008 even though the first flight is now scheduled for the second quarter of 2009. The production slots were all sold out for several years in the future. Before the recession brought on some cancellations they would have liked to be able to build the airplane faster but many factors control their rate of production. The same problem will face the builders of the hardware for solar power satellites.

• • • • • • • • • • • • •

One of the unique characteristics of solar power satellites is the how the

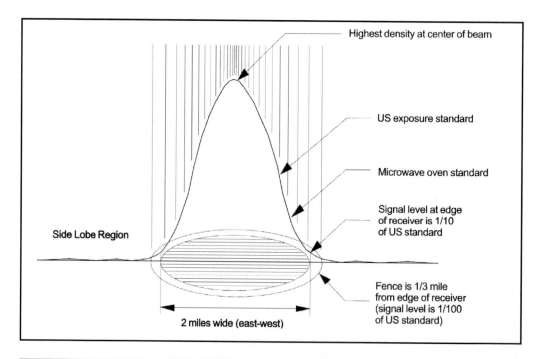

Figure 9 Profile of the energy beam density from the book SUN POWER

energy is delivered to the earth. As I have mentioned it will come to the earth via wireless power transmission beams. Initiated by the phased array transmitters on the satellites, the beam arrives on the earth at a rectifying receiving antenna. There the radio frequency is rectified (changed) into direct current electricity that is then processed into alternating current electricity and then is fed into a distribution grid. One question that is often raised is how safe is this energy beam?

As you can see the energy profile is in a bell shape with the maximum energy density at the center of the beam. This illustration is for a beam that delivers 1,000 megawatts to the earth. At the edge of the receiver the energy density is down to 1 watt per square meter which is the lowest level that is economical to recover. Also shown is the US exposure standard and the allowable leakage standard for microwave ovens.

A 1,000 megawatts is probably the smallest satellite that is practical if it uses a radio frequency wireless power transmitter, as the economics favor the larger satellites. Also increasing the energy density as it reaches the earth to the maximum safe level would keep the size and cost of the rectenna to a minimum.

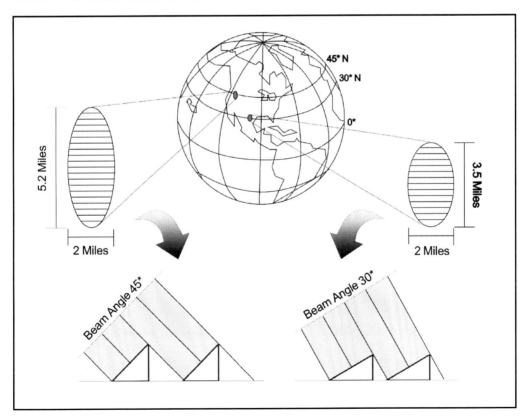

Figure 10 Beam angle as it reaches the Earth from orbit. From the book SUN POWER

The size of the rectenna for a satellite delivering a 1,000 megawatts in the beam will be in the neighborhood of 2 to 3 square miles in area depending on what energy density is selected. The beam will be circular but as most countries are north of the equator with fewer south, the beam will be coming from geosynchronous orbit at an angle. The angle from local vertical is the same as the latitude of the rectenna. This means that the spot on the ground is typically an oval.

The land area required for the rectenna is quite large, but is much less than that required for the mining of coal for the same amount of energy. The rectennas could be owned by the country receiving the energy or they could be owned by the companies that own the satellite. In either case their construction would generate jobs in the receiving countries.

• • • • • • • • • • • •

One dramatic effect on industry would be the requirement of moving the solar cell manufacturing industry from production rates of a few hundred megawatts annually to many gigawatts. It is an industry that already exists, but the requirement for solar power satellites will create a massive number of new jobs. On the other hand are the industries that will disappear as they are displaced by energy from space. It won't be easy, particularly for the older workers who will have to be retrained, but there will be many new industries that will need the workers.

The nature of jobs available in the new industries will be dramatically different. In the energy industry today many of the jobs are associated with the extraction of fuel from the earth. This is true of coal mining, oil rig operations, oil exploration, and natural gas exploration and drilling. With solar power satellites there will be no jobs associated with extraction of fuel. The sun is a fully automated energy source. There will be jobs in construction of the satellite hardware and assembly, fabrication of the earth receiver hardware and installation, and the transportation systems to carry the hardware to space. In addition, as the transition in energy source is made there will be an accompanying change in all the systems currently using fossil fuels, such as automobiles and home heating. The pace of change will be partially controlled by how quickly the work force can change to satisfy the demands of the new industries. The transition could be fairly smooth if existing companies were to recognize what is going to happen and plan for the changes that will take place. This is not likely to happen however, as large mature companies have built themselves around established concepts that have been successful in the past and they are reluctant to believe that they won't work in the future. They will try to stay on course until they have passed the point of no return. New innovative entrepreneurs will see the opportunities and seize the initiative as they have in the software industry and the internet.

There will continue to be jobs associated with the distribution of electrical energy as there is today. The difference will be the great expansion in the use of electricity as it takes over the tasks of heating and production of hydrogen and the desalination of sea water. The satellites will be automated, but it will still be necessary to monitor their operation. They will also require periodic maintenance and replacement of the working fluid for station keeping and attitude control.

The changes made in the transportation industry will be significant, as mass transit systems will be greatly expanded. Private automobiles will change from hydrocarbon fuels to hydrogen power or batteries and later on to the electrification of the roadways. All of this will mean new opportunities. There have been major changes in our societies in the past, but none as big as this that will effect so many people. You may think, "I am not willing to make the change." If that were to be the decision of the majority of earth's people we will be subjected to the spiraling collapse of human society as we continue to pollute our atmosphere, watch the seas overflow our coastal cities and low lying lands. We will see our farm lands turn to deserts, and watch as billions of people die of starvation while we desperately search for the last drops of oil. This may sound far fetched, but history is full of examples of societies that made decisions that led to their ultimate collapse and disappearance. We now have the opportunity to make the decisions that can preserve our world and support humanity into the future.

• • • • • • • • • • • •

I am an optimist and believe we will follow the path of developing a replacement for oil that will stop our polluting of the atmosphere. If we follow the development approach of an industry/government partnership described in Chapter 9, in my congressional testimony in 2000, the first satellite will be built for the Bonneville Power Administration or another government owned utility. It will be the proving grounds for all the critical elements of the required infrastructure and assembly of the system and then its operation. Undoubtedly some problems will arise along the way as there is in any new development, but also lessons will be learned on what works and what doesn't. This first satellite will be the start of the learning curve. In any production operation there is a learning curve that goes on as the product is built. This results in each subsequent product being built at less cost and a faster rate than the one before it. Since all the satellites will be basically the same, this learning curve will apply to them, so as time goes by the cost and assembly time will go down. As time passes there will also be technology improvements that can further reduce their costs.

With the first satellite delivering electrical power to the earth the demand for satellites will blossom. If the United States is the first country to develop the satellites, the question will be, Who will be first in line for the next satellites? I would expect that China and India may be knocking on the door because of their great need for energy. By the time the first satellite comes on line the United States will be suffering from shortages of oil and dramatically escalating prices for gas and diesel. By then global warming and its potential damage to our existence will be evident to most people so the pressure will increase to stop burning fossil fuels. These pressures will be world-wide, requiring international negotiations and agreements to avoid conflicts. The ideal situation is if agreements can be reached to establish a priority list for who will get the satellites first. This isn't a question of who owns the satellites, but rather where the power will be delivered.

The likely model will be similar to the existing energy supply industry. Some companies will build the component parts to be supplied to businesses that assemble the satellites who in turn will sell them to utilities that will own the satellites and sell the power. The receivers on the ground would be built by other utilities or countries that can then distribute the power to the local users. Because of the unique capabilities of the satellites to be located over any location, this opens the potential export of the energy anywhere on earth. In the process of assembling the satellites new industries are created. There will be businesses that develop and build the transportation systems. These could be sold to transportation companies that move the hardware to space. All of the cost will ultimately be paid for by the sale of electric energy to the consumers. This world-wide market is already enormous. But it will grow to several times the current level as energy from space replaces fossil fuels. The business opportunities will be huge during this transition period.

The development of solar power satellites starting during the recession that is sweeping the world in 2009 could be the trigger that revitalizes the economy of the world. India's economy was growing at nearly 9% a year, but this growth cannot be maintained without energy. They desperately need new energy sources. India is only one of the countries in need. China, because of their great demand for energy and the severe atmospheric pollution they are experiencing, must stop building coal fired power plants and start the process of shutting down the ones they have. It will allow their economy to continue to grow and bring an improved standard of living to all of their people. These countries also have a great need for fresh water. With ample nonpolluting electricity they can turn to desalination of sea water to increase their supply. Japan, without its own source of energy can now look to space for clean unlimited supplies. The United States, by far the highest user of energy on earth, can turn from coal and natural gas as their source of electricity to energy from space. The transportation industry can move away from gas and diesel to electric power or hydrogen made from nonpolluting electricity from space.

The growth of solar energy from space in Europe will probably be slower than in the United States because of the extensive use of nuclear power and wind generation. The pace of conversion will depend to a great degree on international awareness and agreements on stopping global warming and the price of oil. If the conversion can be made quickly enough the demand for oil will fall and so will its price. The danger here is if the price at the gas pump fall the resolve to stop burning fossils fuels could be weakened by the desire to go back to gas powered vehicles. It is inevitable that the oil industry will fight this change with all their resources and they are very powerful. If they are really smart they will jump on the bandwagon and lead the way into the new industry, thus saving their precious oil for greater needs.

The use of energy from space will eventually move on to the underdeveloped nations. The process will be slow. Even though the cost of electricity will be low, it does require the installation of the receiver and a distribution grid. These are capital items that will require investment. They don't have to pay for the satellite however, only the energy that comes down to them. Its cost will be less than any other source. What it will do is provide the energy they need to increase their productivity and ultimately their standard of living. How quickly this can happen is uncertain, but it can give them a chance to better the lives of their people.

• • • • • • • • • • • •

Energy from space creates a dilemma for the Middle East and the other major oil producers. How are they going to react when their source of income starts to dry up? I do not have a crystal ball, but the inevitable conclusion is that it will not be a good time for them. A lot will depend on how seriously the world addresses global warming. Worldwide efforts to reduce carbon dioxide emissions in the atmosphere will cause the demand for oil to fall faster than the rate of depletion. Prices will fall and the oil producing nation's economies will shrink. Nearly all the Middle East nations depend totally on oil revenue for income. They do not have industries that can compete in the world market without oil. The possible exception is Dubai that has developed a strong, investment, business, and tourist basis. However, the world recession is hitting Dubai very hard, so they are suffering along with the rest of the world. Others will see their fortunes disappear as their income shrinks with dwindling sales of oil. What will happen in those countries is hard to forecast, but it will lead to serious economic dislocation. They will be in a unique position since they will still have oil, but will not be able to utilize it to produce income. It is uncertain if they will recognize the need to develop other industries to replace oil income in time to avoid disaster.

One of the benefits of their loss of oil revenue will be the fact that they will not have the funds to finance the purchase and distribution of armaments for the radical governments and terrorist groups in the area. They will be forced to use what they have for food and the essentials of life to maintain their people while they develop other sources of income. This will not be a sudden change, because oil will still be used to some extent, but the day will come when most of the nations of the earth will come together and agree to ban most uses of hydrocarbon fuels. Even though the oil producing nations won't agree to the ban, they will be unable to change the reality of their market place being closed.

The situation of the Middle East is a dichotomy. It was the cradle of development for our civilized world, but as technology raised the level of development of the rest of the world, the Middle East languished in their ancient ways. They did not participate in the surge of development brought on by the era of coal and the industrial revolution. When the era of oil burst on the scene at the beginning of the 20th century they still did not participate in the benefits it provided. It was only later that the oil companies of the West came into their lands and opened their oil fields. This began the stream of wealth that followed. But unlike the rest of the world that used the energy to develop industries and evolve their economies into dynamic forces in the expanding world, they remained in the past and ignored the industrial revolution. Even after they expelled the Western oil companies they only used the oil as a source of income for the few and not as a tool to build their nations. They even brought in laborers from other countries to do most of the work. But the increased prosperity did provided the base for exploding populations. It is ironic that this area of the world which has had the greatest riches of nature's resource has failed to use that treasure to build a base for the future. They are simply stripping their land of its resources and gaining no long term benefits, expecting oil to provide them with income forever. The end of the era of oil will bring great suffering for their people if they do not change their ways. The desert land will not be able to support their greatly increased numbers. The area is a ticking time bomb. What will it be like as they face a future with no source of income, a surging population, lack of water, and a desert with no way to even grow their food?

As the era of oil wanes, the fourth era: solar energy from space will be in full swing. It will not be the only energy source, but it will be the primary source. Other sources such as nuclear, wind, terrestrial solar, bio-diesel, ethanol, ocean current, and others will still contribute to the energy pool. By this time the cost of energy from space will be very low. There are three fundamental reasons why this will happen. First of all the technology advancements that will raise the efficiency of the system will in turn decrease the size of the satellites as well as reduce the cost of space transportation. Second, the vast numbers of identical parts needed for the satellites will provide the basis of ever increasing production efficiency and thus reduced cost. And third, as the capital cost of the satellites is paid off, the cost of electricity produced by them is then related to the cost of maintenance, operations, and profit. This means the cost of electricity from these satellites could be less than 2 cents a kilowatt hour. It is likely that as the fourth era matures, the price of energy will fall so that even the poor nations on earth will be able to buy electrical energy to provide for their needs and to grow their standard of living.

The vast amount of energy flowing to the earth will be clean without any atmospheric pollution. It will be available as long as the sun shines. For the earth it is the ultimate energy source, as it has been from the beginning. Solar power satellites are just the direct conversion system that makes the energy useful for us now and into the future.

• • • • • • • • • • • • •

14   The Plugged-in Earth

First it was wood, but wood had its limits. Many parts of the world were deforested as people used the only energy source available to them. Many societies disappeared as a result. Others were struggling in poverty. Coal moved in to fill the void in many parts of the world and elevated those societies to a much higher level of development. It was during the era of coal that electricity started to emerge as an energy form. Then oil brought us into the modern level of today's highly mobile world-spanning society. But it by itself could not power our industry, communication, and computerized technology of today. That takes electricity.

As we move from the era of oil into the era of energy from space we will be changing nearly all of our energy to electricity as the fundamental source. In this chapter I plan to discuss how this can be done. The very first town that was built to be all electric was part of the project of building Grand Coulee Dam in the 1930's. It was a model town where all the engineers, managers and many of the workers lived. The houses had no chimneys and all the heating, lighting, and cooking was done with electricity. This model city was built over 70 years ago to show how electricity could provide all the energy needs for our homes. It is still part of the residential community at Grand Coulee Dam.

Fortunately a great many of our current industries and our homes also use electricity as their energy source. In many cases where oil or natural gas is now used there are electrical systems already developed and used by others for the same purpose. An example of this is home heating. Many homes are heated with electricity, while others use oil, natural gas or propane. Electric heating takes many forms. Our house uses electric baseboard heaters in most rooms, with electric wall heaters in one room and in our bathrooms. Other electric heat systems use electric furnaces or heat pumps. The most efficient electric heat is provided by heat pumps which are like air conditioners but instead of cooling the air they raise the temperature by the use of compressors. They can use outside air as a source of heat or more efficiently use fluid that flows through pipes underground where the temperatures in the winter are higher. Heat pump systems will likely become very common and replace many oil or gas furnaces. Replacing heating systems is relatively easy because so many homes already use electric heat, particularly in areas that have low-cost hydroelectric power. As the price of oil, natural gas and propane rises with dwindling supplies, people will naturally look to electricity because its cost will be lower as solar power satellites are deployed. The primary issue will be the cost of conversion. It will take some time for the reduced cost of energy to pay for the capital cost, but it will soon become obvious that the lower cost of electricity will pay for the conversion in a relatively short time. With new construction it will be easy to incorporate electric heat. Regulations will probably be put in place that prohibit the use of fossil fuels for new construction in order to help in the battle against global warming. This will probably be extended to all heating systems over a programmed time schedule. This happened in England shortly after World War II when they banned the burning of coal in homes in London because of the severe atmospheric pollution.

• • • • • • • • • • • •

The conversion to electricity from fossil fuels will depend on eliminating fossil fuel burning power plants and replacing them with solar power satellites. Undoubtedly the first satellites will be used to fill growing demand. The greatest need is in the developing nations such as India and China where they are desperately adding coal fired plants and nuclear plants. So in the move to reduce atmospheric emissions of carbon dioxide the logical first step is for them to obtain energy from space instead of from more coal fired plants. Nuclear plants do not emit carbon dioxide so may remain in the future power mix to generate electricity, though it will be desirable to phase them out in order to eliminate the problems of nuclear waste as more satellites come on line. As production and deployment of the satellites expand they will bring nonpolluting energy to the rest of the world and the conversion

to electricity will be underway in the entire world including the United States. The first use here will be to meet growing demand. Secondly, the elimination of coal fired plants and then oil and gas fired plants. While this is going on, other renewable sources such as wind and terrestrial solar will be adding to the mix of electrical energy sources.

• • • • • • • • • • • •

The replacement of fossil fuel power plants and the conversion of heating systems to electricity will be the easy part of the transition from fossil fuels. The hard part will be the conversion from oil based fuels to electricity and nonpolluting fuels for our transportation systems. The first step will be to significantly increase the mileage requirements of our automobiles. There is no fundamental reason this cannot be done, except for the resistance of people who want big SUVs and pickups. The mileage capabilities of vehicles in Europe and Asia are already twice as high as the United States. This change alone would make a huge difference in the amount of oil burned. It may be that as we pass peak world oil production the resulting cost increases will be so high that people will have no choice. They simply won't be able to afford their low mileage vehicles. Another enticement could be a major tax on fuels to raise the cost and force conservation. It is unfortunate that recent administrations leading this country have been so blind to the problems we are going to face that they did not start years ago to require high mileage vehicles or to impose much higher taxes on fuels as did the rest of the world. The excuse of damaging the economy is hollow rhetoric. Just look at the fact that Japanese automakers are flourishing while United States automakers are on the financial rocks.

The next transitional step is already underway as hybrid vehicles are becoming common sights on our highways. I have friends with Toyota Prius cars that have great fun seeing how high a mileage they can achieve. With careful driving in city commutes they often get 70 miles to the gallon. These are comfortable four passenger cars. When you drive the highways and streets of our city and freeways most vehicles have only one person. It is ludicrous that they are often in a big SUV. The hybrid vehicles being sold by United States manufacturers tend not to be high mileage vehicles, but rather large vehicles that use the electrical system to achieve higher performance rather than better mileage. United States automobile manufactures were begging the government for money to keep them out of bankruptcy as sales of their vehicles plummeted. Unless they see a way to restructure themselves they will simply accelerate their low mileage vehicles to their demise. I can see the picture now as a fleet of SUVs drive over the horizon into the sunset.

Hybrids offer a good solid step towards reducing gas usage. The next step along this path is plug-in hybrids. These are vehicles that use the technology of today's hybrids, but add the capability to charge their batteries from an electrical outlet to supplement their onboard engine. With improved batteries the plug-in hybrids will be getting 100 to 150 miles to the gallon. A company in California is already converting new Prius cars to plug-in units and replacing the existing batteries with lithium ion batteries. The warranties are no longer honored by Toyota, but the cars get spectacular mileage. As technology advances, plug-in hybrids will become common. They are essentially electric cars with an engine that eliminates the danger of becoming stranded when the batteries are depleted. Another potential modification would add solar cells to the car that would supplement charging the battery. Hybrids will undoubtedly be a big part of our future as we make the transition away from oil. It will be interesting to see if the United States automakers can make this step as we move into the future. Their future will depend on it unless they can make it to the next step first, which will be all electric or hydrogen fueled cars, based on using hydrogen made from water with electricity from space.

Let me tell you the story of the electric car. General Motors started this step to electric cars powered by batteries several years ago with their EV-1. It was developed to meet the requirement established by California that 2% of all cars sold in the state be emission free starting in 1998. Many of these cars were built and leased to users. Their initial range was quite short, but with improved metal hydride batteries their range was significantly increased. The people that used these cars loved them. They had good performance and met most of the range requirements for local driving. However, opposition from the oil companies and the auto-mobile manufacturers surfaced and they financed ad campaigns against the mandate and brought law suits against California to prevent the mandate. This was joined by the Federal Government so California caved in under the pressure and eliminated the requirement. General Motors then recalled all the electric cars over the strong objection of the users. They refused to sell the cars to their users and forced all of them to be returned. General Motors then proceeded to destroy them all. Unbelievable!

It seemed to have been the first act of suicide for General Motors. Only time will tell if they can do an about face and survive. The bailout by the government at the end of 2008 to pre-vent their bankruptcy may just delay the inevitable. In March of 2009 they are teetering on the edge of bankruptcy again.

Ironically after their massive destruction of a promising new con-cept in the EV-1 in 2003 they announced in early 2007 that they had awarded contracts to two firms for the development of electric car batteries. These contracts are for developing lithium ion batteries for a new plug-in hybrid vehicle. If they are successful, it would be the next giant leap into the future for electric vehicles. By winter of 2007 General Motors was showing off the prototype of their plug-in hybrid, the Chevy Volt. It is a very attractive automobile that uses the new lithium ion batteries and will probably be the standard bearer of General Motor's effort to recover their domi-nate position in the industry. One won-ders if the same fate could be lurking in the future if they are again successful.

The experiment with EV-1 did prove that electric cars with battery power can replace a vast majority of vehicles that are used for normal local driving. Nearly all commuter driving is for ranges well within the capability of existing battery technology. If you add recharging capability at the work place, most commuter traffic could be accomplished with battery powered vehicles. However, they must ultimate-ly receive the electricity for charging from a nonpolluting source. The Cali-fornia mandate approach should be revived and made nation-wide to force the transition away from oil based fuels that add carbon dioxide and other pol-lutants to the atmosphere. As this occurs, battery technology will improve. For example it is lithium ion batteries that now power our laptop computers and cell phones. The Chevy Volt will have a range of 40 miles using batteries alone. The potential is for increases in range of electric auto-mobiles to several hundred miles. There are questions about the safety of lithium ion batteries, but with technology they will continue to evolve with the large financial incentive of a huge market which will undoubtedly accelerate solving the problems. Another example of new development is lithium/air batteries. They have the potential to greatly improve on the weight and power output of lithium ion batteries.

In the March, 2, 2009 issue of Newsweek magazine there was an article, titled, **To Pack a Real Punch**. It was an interview by Newsweek's, Fareed Zakaria with Alex Molinaroli, president of power solutions at Johnson Controls, which will supply the battery for Ford's first plug in electric car. The questions and answers given in this article explored the potential of battery powered cars and the need for the United States companies to step up to the challenge of developing and manufacturing batteries here in the United States because of the great business potential that is out there in the future as we move away from gasoline powered vehicles. Alex Molinaroli was confident that they were up to the challenge of developing the lithium based batteries that were required and indicated that they were supplying Daimiler's first vehicle lithium battery and also BMW's first lithium battery will be their's. He indicated that the consumer battery industry had moved off shore 15 to 20 years ago. They want to put manufacturing back in the United States, and that government support would give them that opportunity.

Ford is the only one of the big three

US automobile manufactures that is not asking for a government bailout nor is on the verge of bankruptcy. Their turning to a plug in hybrid may give them another massive boost in the competitive world of the future.

Electric cars can certainly solve the problem of local driving. The issue of long distance cross country driving is a different problem. It tends to be unique to large nations such as the United States. Stops to recharge batteries could be a nuisance. One potential way to increase range is the installation of high efficiency solar cells on the upper surface of the vehicles. Automobiles powered by solar cells race across Australia at high speeds. These are very light aerodynamically designed cars, which do not represent family automobiles, but they show that there can be a significant added energy source for family cars. Another option is exchange of battery packs at cross country service stations. This could be done in minutes with proper design of the automobile and service equipment. Since most households own multiple vehicles it would be practical to have one or two electric cars and one, hydrogen powered for long range driving. Another example of the technology improvements in batteries, DeWALT battery powered hand tools are available with modified lithium ion batteries that can be recharged to 80% of capacity in five minutes. With batteries of this type in cars the issue of range for battery power vehicles may be solved.

• • • • • • • • • • • • •

Hydrogen is potentially the fuel of the future that can be made from water by electrolysis when the electricity comes from solar power satellites. The Bush administration's forecast that hydrogen is the fuel of the future rang hollow since their source of hydrogen would be fossil fuels. It does little good to convert to hydrogen if the source is natural gas or some other fossil fuel. The extraction still leaves pollutants. However, when it is extracted from water it becomes a very desirable fuel as the product of combustion is water. In Iceland they use hydrogen as fuel that has been obtained from water using electricity generated with terrestrial solar cells or nonpolluting geothermal energy. The difficulty comes in how to store it for use in an automobile. It is normally a very lightweight gas at room temperature and very easily ignited if oxygen is present. A hydrogen flame is nearly invisible in daylight. If you have ever watched a Space Shuttle launch you have seen the tremendous plum of fire and exhaust gases ejected by the solid rocket boosters, but it does not look like the engines on the orbiter are even running because you cannot see the flame coming from the three hydrogen burning engines. They are firing and burning large quantities of hydrogen and oxygen that comes out of the nozzles as super heated water vapor that is transparent.

There are three methods of storing hydrogen at the present time. The first is to pressurize the gas to high pressure and store it in a high pressure tank. 10,000 pounds per square inch (psi) is not uncommon. Even at that very high pressure it takes a fairly large tank to store a significant amount of hydrogen because it is so light. It has only half the weight of helium. Its energy capacity per pound is the highest of all elements, but because of low unit weight it requires large volumes to contain it. Pressurized tanks are heavy and at 10,000 psi fairly dangerous. The second method of storage is to liquefy it by reducing its temperature to minus 423 degrees Fahrenheit. This is the way it is handled in the external tank for the Space Shuttle and other hydrogen fueled rocket systems. Even in liquid form it only weighs 4 pounds per cubic foot, so the volume is pretty high and it takes really good insulation to keep it liquefied. Liquid hydrogen is probably not the way you want to store it in your automobile.

The third way of storing it is much more user friendly and will probably become the most common method. In this method you would have a fuel tank filled with metal hydrides. They have the unique ability to absorb hydrogen. Some alloys can actually absorb more hydrogen in a given volume than if the hydrogen was liquefied. The pressure that is required is relatively low. As the hydrogen is used the hydrides can be made to release it. The weight of the metal hydrides is not light, but it is a system that allows driving up to a hydrogen filling station, attaching a hose to a filler valve on your tank and "filling her up." This kind of system is in operation in Iceland. The same company that developed hydride batteries is working on advanced alloys for storing hydrogen.

Storage is probably the biggest difficulty facing the conversion to hydrogen as a major automobile fuel, as its other characteristics are so desirable.

How hydrogen will be used to provide energy to power the wheels of our automobiles is still uncertain. Fuel cells using hydrogen and oxygen to generate electricity have been in use for various functions for many years and automobile manufactures have used them in experimental vehicles of various types. Hydrogen can also be used in internal combustion engines. Whatever system turns out to be best will evolve when market demand for hydrogen fueled vehicles develops as oil based fuels are replaced. In addition the infrastructure necessary to support hydrogen fueled vehicles will have to be developed in parallel with vehicle development. One thing that can make it fairly easy is the ability of a hydrogen service station to produce its own hydrogen from water using electricity from the grid. It will not be necessary to transport the hydrogen from place to place as it is for gas and diesel.

Honda is currently leading the way with hydrogen fueled vehicles. They are developing a hydrogen fuel cell powered car that will soon be available in some areas where hydrogen fueling sources are available. There are a few in California. They will not be selling cars to the users, but will lease them in a similar way that General Motors did with the EV-1. Honda felt they missed out on the hybrid move to Toyota and is going to try and lead the way into hydrogen powered cars. It will be an interesting experiment to watch.

• • • • • • • • • • • • •

Another potential option for long distance travel by private automobiles would be the electrification of the highway system. There have been experimental systems that transfer electrical energy by induction pads buried in the roadway to a passing vehicle without actual contact. If you have a Sonicare tooth brush you know what I mean. You simply place it in its holder and the battery is kept charged by induction from the holder to the tooth brush handle that contains the battery. There is no electrical contact. This approach could possibly be used to pass energy in the roadway to the automobiles passing over it. With this approach battery powered cars could have cross continent range with no pollution or stops to the refill the fuel tank.

Electrifying the interstate highway system might be one of the developments that we can look forward to watching as technology evolves. It is the logical step to making battery cars universal. There will be a new competition between electric cars and hydrogen fueled cars. Either one can solve the problem of eliminating carbon dioxide emissions, as long as the charging electricity and the energy to free hydrogen from water comes from nonpolluting solar power satellites or other nonpolluting sources.

• • • • • • • • • • • •

The conversion of our personal transportation vehicles is only one aspect of our total transportation industry, even though it is a very large part of it. The trucking industry is another major part. Trying to power a long haul eighteen wheeler with batteries is probably not going to work very well. Even, trying to use hybrid designs for long haul will probably not help much. However, for city delivery, hybrids may be very effective where much of the driving is stop and go and braking energy can be fed back into the batteries. There are already some hybrid heavy haul trucks reaching the market for local use. The most likely approach is to power the long haul trucks with hydrogen. They have the ability to carry bulkier containers for the hydrogen and it will probably be practical for them to use liquid hydrogen. They could even have refrigeration systems incorporated into the truck design that would maintain the low liquid hydrogen temperature. It is likely that they would have internal combustion engines burning hydrogen, at least until fuel cell design could economically supply the power.

The other likely change that would occur in long haul freight transport would be the conversion and expansion of the railroad system from diesel to electric power. Much of the rest of the world uses electricity to power their trains. The United States now uses electricity to power the light rail systems that carry passengers in and around our cities. The conversion of long haul rail lines can be made gradually as the railroads are a very efficient mode of heavy transportation. They will probably be expanded to take over part of the long haul trucking business. The degree to how much will depend on how efficient and economical the trucking industry becomes as it transitions to hydrogen power. It is possible the electrification of the highway system could be carried over to long haul trucks. The railroad system benefits from the fact that they always travel on the same controlled right-of-ways so they will be easy to electrify.

• • • • • • • • • • • •

Expansion of our mass transit system is an important part of moving away from fossil fuels. The light rail systems currently in operation are already electrified and there are still some buses and trolley systems that are electrified. Seattle has joined other cities in the use of hybrid buses that double the mileage of conventional diesel buses. Expansion of mass transit systems has several benefits. They are efficient in carrying large numbers of people. They reduce crowding on streets and highways. They provide convenience and low-cost transportation in city environments. They can significantly reduce energy use.

Europe and Japan lead in the use of mass transit. As the population growth continues in the United States, it will become increasingly more important if we are to avoid the massive traffic jams that choke our cities today. The expansion must be with electric powered systems. This means that light rail or monorail systems should eventually replace busses, unless bus systems can be developed using batteries or fly-wheels and induction charging systems are developed. The use of hybrid busses is a step in this direction.

• • • • • • • • • • • •

So we have been looking at transportation systems that are fairly easy to convert away from oil fueled systems. There are a couple more that are not so easy. They are ships and airplanes. First let us consider ships as they have a couple of ways they can go. One is for them to become nuclear powered. Today several navies of the world have nuclear powered submarines and/or aircraft carriers that are nuclear powered. One of the first nuclear powered ships was a cargo vessel built by the United States to show the world that nuclear power could ultimately power the commerce of the seas. Unfortunately, timing and world tensions worked against its success and it was ultimately scrapped before it could prove its worth. Nuclear power could become the driving force of future commercial vessels, particularly larger container ships. The giant oil tankers of today may disappear when their source of cargo disappears. The expansion of nuclear energy to the relatively unreg-

ulated ships that ply the seas of the world will not be an easy development. Concerns over nuclear proliferation will undoubtedly restrict its evolution.

An easier transition would be to hydrogen. As it is with the trucking industry volume is not a big problem. Liquid hydrogen could be the best way to carry the fuel. However, safety will be a primary concern. One of the problems with hydrogen is it is a very difficult gas to contain because of the small size of the molecule. It is also easy to ignite so any small leak is likely to be ignited before the gas can disperse. If the area where leaks might occur is well lighted, or in sunshine, you cannot see that it is burning. In the early days of rockets using liquid hydrogen the technicians would sweep the pipe joints with a broom. If it burst into flame they knew there was a leak and the hydrogen was burning. That is one of the other characteristics of hydrogen, it is extremely flammable, but as a result it rarely explodes like gasoline. You may have seen old photos or news reels of the burning of the Hindenburg dirigible. The hydrogen burned, but did not explode. The smoke and flames visible in the photos is from the burning fabric not from the hydrogen that filled the dirigible. Hydrogen was used by the Germans because the United States, with the only source of helium at the time, refused to sell them the nonflammable helium.

There is an interesting new development in Long Beach, California. Foss, one of the largest tug boat companies in the country, is joining with the Port of Long Beach to give the shipping industry an ecological makeover, by adding an electrical hybrid system to their tug's powerful diesel engines.

"It should have a profound impact on tug technology in the decades ahead." Port of Los Angeles spokesman Arley Baker said. "It's a huge step forward." The first boat could be in operation by 2008 in the Los Angeles area, which is the nation's largest port complex.

The hybrid design used by Foss is similar to the technology used by hybrid cars. It also follows the diesel hybrids used in railroad vehicles. The tugs would still have diesel engines to provide the primary horsepower for handling ships and barges, but for idling and less strenuous tasks, they would use batteries, supplemented by diesel generators for power.

The pilot project is part of a project to cut pollution in the country's two busiest container ports. The use of the tugs could be expanded if they meet expectations in reducing emissions. Foss hopes to offer conversion systems to retrofit older tugs.

• • • • • • • • • • • •

So far we have found a way to replace oil in all of our transportation systems, except airplanes. They are by far the toughest problem we face as we move beyond the era of oil. There is no obvious replacement for oil based fuels for airplanes as we know them today, except biomass fuels. The best approach that I can suggest is for the world to agree to prioritize the use of a limited amount of oil based fuels for use in the airline and air cargo industry. With the conversion of all other energy systems to electricity and hydrogen extracted from water with electricity the use of fossil fuels will be dramatically reduced so that the amount used by aircraft will be small enough that the world environment can support the reduced levels needed by airplanes. This will give the inventive mind of mankind time to find other solutions. Current airplane technology would not support the conversion to hydrogen, because of its high volume requirements.

One exception may be for hypersonic flight on long range flights. When I was working on Boeing's National Aero-Space Plane program our propulsion engineers developed a theoretical curve of range versus speed. The range was fairly constant as speed was increased from a couple of hundred miles per hour until mach one was approached. Then it dropped significantly. However, as the mach numbers rose, the range started to increase. As it moved into hypersonic speeds the range got longer and longer. What this told us was if we could build air breathing engines that could go to really high mach numbers we could have long range ultra fast airplanes that would use much less energy than our current airplanes. These airplanes would actually go out of the atmosphere for part of their flight and have a ballistic glide to their destination. If this is true the day could come when it will be less than an hour's flight to any place on earth. These vehicles would use hypersonic scramjets burning hydrogen. They would have to use hydrogen as it is the only fuel that has a fast enough burn rate to satisfy the requirements of the engine. Since the time I worked on the National Aero-Space Plane there has been significant progress in developing hypersonic scramjets.

The other option, as I mentioned, is bio-fuel. The issue here is; what is the source of the biological fuel? If it is something that uses large areas of land that could be used for growing food it is probably not a good idea. However, algae that is grown in vats on vacant land such as under solar power satellite rectennas, could be a very good source. There is work going on to tailor the algae to provide just the right kind of fuel. It is a fuel the has great potential.

• • • • • • • • • • • •

Much of the world is short of fresh water. With ample, low-cost, nonpolluting electricity available, it is practical to power reverse osmosis desalination systems to convert sea water to fresh water. The reverse osmosis process can also purify brackish, or contaminated water. It would allow the means to recycle waste water where there is no salt water source available and fresh water is limited. It could

mean the difference of an area supporting life or being uninhabitable.

One of the unique characteristics of solar power satellites is that they will be producing energy at maximum capacity at all times. So as normal electrical demand fluctuates during the day night cycle, the energy not used is lost. However, this could also allow us to use the excess power to extract hydrogen from water during the low demand periods as well as desalinate sea water. That way all the energy is used and none wasted. It's a little like a free lunch.

• • • • • • • • • • • •

As you can see the transition from oil to electricity from space is not only possible, it will also be practical. It will be a transition of massive proportions that will present countless opportunities to counteract the difficulties that change always brings. It will provide for a world that can have a pollution free atmosphere and eliminate the need to tear up the earth in search of its resources. It can provide the means for supplying fresh water and hydrogen fuel and heat our homes. It will power our cars and our industries. Electricity is the highest form of energy we have. And with clever rectenna design, we can also grow our food and help feed the world's hungry people or grow algae to make bio-fuel.

The solar power satellite reference system developed in 1977-80, was based on the technology of that period. Even 30 years ago it provided a system that was highly efficient and had promising energy costs. Since then there have been great strides in technology that will make significant improvements.

The most significant development has been in solar cells. The original design used single crystal silicon solar cells. These cells were much thinner than those used in terrestrial applications in order to save weight and material. Their thickness was only 2 millimeters, compared to about 7 or 8 millimeters for other cells of the day. The thinness of the cells also helped in making them more resistant to space radiation degradation. The reason they did not degrade as fast as the thicker cells was because the damage occurs where the high energy particles are stopped in the cell material. With the thinner cells more of the radiation particles pass through the cells without being stopped, and therefore do not damage the cells. The cells were also thin enough to be fairly flexible. Their efficiency was 16.5% of the sunlight in space. The number of options today is greatly expanded due to the development that has occurred over the years. This includes solar cells with efficiencies in the range of 40% for multi-band cap cells and ultra light thin film cells that are now in production.

The cells that appear to be the wave of the future and ideal for solar power satellites are the thin film cells made from copper indium gallium selenide (CIGS), deposited on metal foil. The first company to use this material in light weight cells was DayStar the company formed by Dr. John Tuttle after he left the National Renewable Energy Laboratory (NREL). Using SIGS technology deposited on titanium foil he was able to deliver 2,000 watts of electricity with only one kilogram of mass. Later on in 2006 Nonosolar announced they have developed a way to apply the material using a printing process on metal foil in an endless roll that is similar to printing a newspaper. They say their cost will be one fifth to one tenth of the cost of silicon cells with similar efficiency. They are building the largest solar cell factory in the world to produce these cells. Other start-up companies are joining the effort as well. Another sign of the growing interest in CIGS was when Shell, one of the largest solar companies in the world, sold its silicon solar business to focus on developing **CIGS**.

Thin film cells will revolutionize the solar cell industry. The thin film cells on metal foil are ideal candidates for solar power satellites. They are light weight, low-cost, and will be very durable in space. Because they are so thin, there will be practically no radiation degradation in space. Nor will there be any degradation from sunlight so their life should be indefinite. This feature is extremely important to solar power satellites.

This is not the only advancement that has been made in solar cells. Spectrolab, a subsidiary of Boeing, has entered the terrestrial solar cell market with a system that uses ultra-high efficiency Gallium Arsenide based multi-junction cells that use concentrators. These concentrators multiply the sunlight on each individual cell.

They achieved the highest efficiency ever with a cell reaching 40.7% efficient with a 200 sun concentrator on December 6, 2006. The average production efficiency for these cells with 500 sun concentrators is 35%. Australia ordered 500,000 cells that will be able to supply 3,500 homes with electric power. Another order is for a 10 megawatt installation from a company in Palo Alto, California. This will supply power to 4,000 homes. The cost of the electricity produced will be about half of the cost from a silicon cell system. Each cell is only one centimeter square but produces 15 watts of electricity in a terrestrial installation with 500 sun concentrators. Since the sunlight in space is nearly 50% more intense than it is on the earth, the output in space would be even greater.

If these cells were used on a solar power satellite, the satellite size could be reduced to less than half the size of the 1980 Reference System. The trades that must be conducted before selecting a solar cell type for the satellites now must consider satellite size, weight, cost of the cells, difficulty of transporting the cells, and the complexity of installation. In all cases the satellite will be oriented to face the sun at all times so this would not be an added problem for using concentrators for the cells

By its very nature, the heart of the solar satellite is the solar cells that generate electricity from sunlight. As a result the advancements made in solar cells has a huge impact on the long term viability of the system.

Another major element of the satellite that has a large impact on the cost, weight, and assembly of the satellite is the structure that supports the cell blankets and maintains its shape. The material selected for the structure during the definition studies was aluminum. It is light weight, easy to fabricate and has extremely long life in space. In the years since the early studies there have been great strides made in using composite structure to achieve low weight and durable long life structure for airplanes. The use of this material has steadily increased over the years.

The ultimate current example is Boeing's 787 Dreamliner which is built primarily of composites. Whether this is the material to use for the satellite structure is still unknown. One of the reasons it was not selected earlier is because it is subject to degradation in the space environment if it does not have a protective coating that is not required for aluminum. The other key issue is ease of manufacture and assembly. There are a couple of options to consider concerning the structure. One is to manufacture component pieces on the earth that can be conveniently packaged for transportation to space and assembled there. The other is to take raw materials to space and do the manufacturing there. Rolls of aluminum could be taken to space to supply the material for an automated beam builder that can produce endless triangular truss beams. Grumman made a prototype of this approach in the late 1970's.

At this time it is not clear what the best approach will be, but what is certain is the technology for automated manufacture and assembly has continued to evolve over the years. Robotic assembly has been used extensively in the automobile industry and there is a solid technology basis for developing the machines that will be required in space for assembly of the satellites. Robotic assembly will be mandatory as men in space suits would not be practical as is apparent from watching the space walks on the International Space Station.

Another interesting approach might be the use of inflatable structure that is folded into compact packages for transportation to space. When in space it could be inflated. A resin impregnated in its fabric would cure after deployment making the material rigid.

The next major element of the main part of the satellite is the wiring to gather the electricity from the solar cells and feed it through the rotary joint to the transmitter. The rotary joint is required to allow the satellite to face the sun at all times while the transmitter faces the earth. Thus the transmitter makes one revolution each day relative to the satellite. The trades for the electricity gathering will consider DC current versus AC and the best voltages to use to minimize weight and problems of static electrical build ups.

The reference design used slip rings for transferring electricity across the rotary joint, but an alternate method could be via a transformer if the electricity distribution was with AC current. Analysis made during the System Definition Studies concluded that the slip ring approach would not be a significant problem as the actual motion was so slow because the transmitting antenna makes only one revolution per day. This means that it would make 36,500 revolutions in 100 years of operation. By comparison a typical DC electric motor using slip rings, running at 1,800 rpm would make 2,664,000 revolutions in one day and they can run for years.

Another important part of the satellite itself is the attitude control system. Ion thrusters powered with electricity from the satellite appear to be clear winners for this task. But one of the sub-trades would be to use the thrusters to maintain the shape of the satellite in addition to its attitude and orbit position. This could be done so the structure could be made very light and flexible. With the computing technology available today that might be a good choice.

The last part of the satellite is the transmitter. It is what makes electrical energy from space possible. The design concept we used on the reference system was patterned after phased array radars, like the DEW Line radars installed in Alaska during the cold war. This design uses a series of sub-arrays covered with slotted wave guides that radiate the radio frequency energy. Each sub-array had a high power klystron that converted the electricity into radio frequency energy. Klystrons are used extensively in high power phased array radars. Steering of the beam is accomplished by controlling the phase of the frequency in relation to the adjacent sub-arrays. In **Sun Power**, I described an alternate approach that was developed by Bill Brown, the inventor of wireless power transmission, which used microwave oven magnetrons. By converting them from oscillators to phased lock amplifiers they are much more efficient and also very low-cost. Because of the smaller output they did not need active cooling. These are still a good option for the transmitter. Raytheon has continued development of energy transmission systems in various programs for the military that are not yet available to the public. They are now working on studies of the satellite system that will probably incorporate new advanced technology so they are confident they

know how to design the transmitters. They may also be able to modify the energy distribution and reduce the diameter of the beam as it reaches the earth. This would allow smaller satellites to be practical or multiple beams from one satellite.

One of the important steps in developing the best system would be a ground demonstration of the wireless power transmission system on a small scale. This could become an ongoing development facility that would be used to test and demonstrate new technology and systems.

• • • • • • • • • • • •

This satellite design concept is based on the Boeing Reference System developed in the 1977 to 1980 time period that follows Peter Glaser's original concept of a satellite in geosynchronous orbit that used solar cells in an array that faced the sun at all times and had an energy transmitter that faced the earth all the time, so there was one revolution each 24 hours between the two elements. This is still the logical and best approach. However, other configurations have been studied or suggested. One was the power tower configuration that was developed during the NASA New Look Studies in the 1995-2001 time period. The emphasis during these studies was to try and find a way to start the program with less initial cost. This included looking at niche markets, alternate orbits, small satellites, systems that could be launched with expendable launchers, and other ideas. However, the resulting configurations do not lend themselves to solving the problem of developing the vast amounts of electricity that the world needs. They do reflect the culture that has evolved in NASA that is more focused on new, cutting edge technology than on operational systems.

Another configuration idea was unveiled at a press conference in Washington DC on October 10, 2007 by John Mankins the former NASA employee who headed up the New Look Studies. His idea was to develop a sandwich structure composed of the energy transmitter with high efficiency solar cells on the back side. The transmitter would face the earth at all times. There would be two giant circular mirror arrays that would be supported on structural booms that allowed them to move as the satellite orbited the earth. They would concentrate the sunlight onto two secondary mirrors that would then reflect it onto the solar cell array with high concentration. The idea was to eliminate the need for a rotary joint and an electrical wire collection system. This configuration, though promising, has not yet had the opportunity to be engineered to identify its capability and commercial viability.

There will be an avalanche of creative configuration ideas that will undoubtedly surface when satellite deployment starts to become real. It will be up to the commercial companies to design and develop the best configuration based on the available technology that will provide the lowest-cost electricity to earth.

● ● ● ● ● ● ● ● ● ● ● ●

The real elephant in the room that must be discussed is the problem of how to transport the satellite hardware to space. It is a crucial part of the system and the reason the satellites have not yet been developed. So it is worth exploring in more detail. I am going to concentrate on the third of the three options. The first is the plan developed by Space Island Group. By modifying the design of the external tanks currently used by the Space Shuttle, they would be taken on to orbit for use as commercial space stations that in time would pay for the launch cost. Their system would eliminate the orbiter and add engines to the external tank. The source of funds to develop the systems is based on selling electricity from solar power satellites. The limiting factor of this concept is the number of launches until the commercial space station market becomes saturated and a new launch system must be developed.

The second option is the Space Elevator which will emerge as the technology is further developed. However, it cannot be done today, whereas the first and third approaches are doable now.

The third option is to develop a new fully reusable heavy lift commercially viable launch vehicle. In the past all the trade studies conducted for the military, NASA, and even commercial space satellite users, always concluded that the lowest cost launcher should be an expendable (throw-a-way) vehicle. This is because the market size was never large enough to justify the expense of developing a fully reusable system, nor was the lift requirement large enough to justify large lift capability to low earth orbit. The original concept for the Space Shuttle was a fully reusable two stage vehicle, but politics, lack of a clear goal, and bureaucratic mismanagement doomed it from the start. The potential market size created by solar power satellites more than justifies the large development investment needed for a new reusable launch system. It must be commercially designed and operated, not an experimental or developmental system.

Over the years since space was opened in 1957 by Sputnik there have been a great number of launch vehicles built and flown by many different nations. The biggest and probably the most famous was the Saturn V moon rocket. During that period the Russians were also trying to build a big launcher to go to the moon. It was actually a little bigger than the Saturn V, but it was never launched successfully and the United States won the moon race. Now the Russians are partners, along with several other nations in the International Space Station. Today commercial launch vehicles are being built by the United States, Europe, Russia, Japan, China, India, and probably other countries are looking into the possibilities. Many military launch systems have also been developed and flown.

Some of the vehicles have returnable capsules and there have been some attempts to develop small reusable launch systems. The one fully reusable system that is going to be in operation in 2010 is Virgin Galactic's Space-ShipTwo. It will only be a sub-orbital system however. The Space Shuttle is the closest to a reusable launch system and has proven that an orbital reusable vehicle can be built.

The criteria for a successful launch system that will have low enough launch costs to deliver satellite hardware to orbit can be summarized in the following few, but very critical requirements. These are the general criteria that drive a large number of sub-criteria that must be met to be successful.

1) It must have a large payload capability, at least 200,000 pounds.

2) It must be able to fly very frequently with a goal of two flights a day for the booster and one flight a day for the orbiter.

3) It will have a payload bay that can handle light weight standard sized shipping containers.

4) The orbiter must be mated to the booster in an arrangement that will eliminate the possibility of ice damage during launch.

The logical launch vehicle for solar power satellite hardware that has the potential to be able to meet these requirements is a two stage fully reusable system. There are options as to what it will look like, but, when all of the requirements are considered it will most likely be a winged booster and a winged orbiter with their shapes being similar to the Space Shuttle orbiter with the orbiter tail mated to the booster nose. The orbiter will use hydrogen for its fuel and the Booster will use hydrocarbon fuel initially. It would be desirable to have the fuel be Liquid Methane, with new hydrogen fueled boosters replacing them at some point after the initial satellites are in operation.

If these criteria are met there is no question that the cost of launching solar power satellite hardware will be low enough that the cost of electricity from the satellites will be competitive with all other sources of energy and when the capital costs are paid off will produce the lowest cost electricity in the world. The big stumbling block is the initial cost of developing the vehicles. When you look at the enormous number of launches that will be required to satisfy the world market for electricity from space, you soon recognize that the cost of the vehicles can be amortized over this huge number of launches, so that is not the problem, it's the up front development costs that must be paid before the market can be opened. This is where the government needs to help.

• • • • • • • • • • •

The last element of the system is the ground rectifying receiving antenna. This is a simple concept that is composed of half wave dipole antenna elements and diodes to rectify the radio frequency energy to direct current electricity. The test of this concept by NASA at Goldstone in 1975 demonstrated 82% efficiency for the rectenna in receiving and converting the radio frequency energy to electricity. Technology improvements since then will raise the receiving efficiency to over 92% and the design can be simplified to reduce the number of diodes by having several antenna elements feed one diode.

In going over the technology advancements that have been made over the years since the completion of the System Definition Studies in 1980, I hope you have reached two conclusions. The first is that even though the satellites would be huge, they are really quite simple in concept and design and the technology in 1980 was good enough to have produced working satellites at that time. Second the technology advancements since then now provide a sound basis to produce satellites that will be even lower-cost. They will be able to provide electrical energy to the earth that is at a lower-cost over the long term than any source except possibly hydroelectric. There is absolutely no technological reason for not proceeding with implementing the fourth energy era, the era of energy from space.

16   **Our Future**

Our future is in great question today with the world economy in tatters. At the same time it gives us the opportunity and incentives to make a dynamic leap forward into a future of prosperity unmatched in the past. It takes a crisis to shock people into action to make the dramatic changes that are required. We are in a crisis today. It is a crisis that is driven by four worldwide problems. The economy has crumbled, we are experiencing the inevitable signs of the depletion of oil, global warming brought on by our burning of fossil fuel is threatening the earth as we know it, and wars over oil hang over our heads.

If we can rally all the people of the world to join forces and solve these problems we would move into a new era of economic growth that would encompass the entire earth. There is a solution! It comes down to energy. If we look to the sun which is our ultimate source of energy we find the answer. If we develop solar power satellites, we would tap the ultimate source directly and provide for bringing low-cost, non polluting energy to the whole world, 24 hours a day. It would open a vast new industry that would bring jobs and prosperity and open entirely new frontiers to explore. Solar power satellites would replace oil as the major energy source for the earth. They would allow us to stop burning fossil fuels and allow the earth to heal. They would eliminate the incentive for wars over oil. They would change our source of energy to the sun. They would start a new wave of economic development that would sweep around the world. They would create another industrial revolution. Our crisis would end and the world would enter a new era, the era of solar energy from space.

As in all great adventures the first step is the hardest. I can clearly remember August the 10th 1987, my birthday. My wife and I had been preparing for nearly twenty years to leave our world behind and go cruising the oceans of the world in our own sailboat. This was the day we were to cast off the lines and go.

Our dream had started a couple of years before men walked on the moon when some friends from my days in the Air Force showed up in New Orleans after a trip down the Mississippi River from Minnesota in their 50 foot Yacht. They took us for a ride on Lake Pontchartrain and since they were going to leave the boat in New Orleans for several weeks they invited us to use the boat while they were gone. The experience of being on a boat like that set in motion a new direction for the rest of our lives.

We found a house in the bayou country north of Lake Pontchartrain with a boat slip in the back yard. It was soon filled with our family Christmas present that year, a 23 foot sailboat. Over the next couple of years we had a wonderful time learning to sail. The dream was building. When I was called back to Seattle to work on the Space Shuttle Definition Studies we had to sell the boat, but the dream persisted. We became the proud owners of a classic 35 foot wooden sailboat built in 1939. It was not long before we were making extensive forays north into Canada. It was during this period that our dream started to take on specific form. We set as our goal to retire early and sail the oceans of the world. We would sell our current boat and find one that was larger and more seaworthy.

After a period of unsuccessful searching, I decided to build our own boat. Seven very difficult years later the result was a beautiful 49 foot classic ketch. We named her FRAM, after Fritjof Nansen's boat that he used in his attempt to reach the North Pole in the 1890's and that was later used by Amundsen when he went to the South Pole. We launched her with great fanfare in the summer of 1978 and it soon became our home. She was seldom at our dock on weekends or vacation time. Puget Sound, the San Juan Islands, and Canada became our play ground as we learned to sail her. The miles added up fast as we put thirty thousand miles under her keel. In the summer of 1986 we did a circumnavigation of Vancouver Island to feel the experience of sailing on the open ocean and big seas.

Finally our date with adventure arrived and it was time to untie the lines and cast off for the oceans of the world. Our only clear destination was Cabo San Lucas, Mexico by Christmas. After that the winds of chance were going to direct us on which way to go. We eventually reached Australia, half way around the world. As we met people from various lands, they would wistfully say, "Oh, I'd love to do what you are doing." I would tell them, "Just untie the lines and go." If only it was that easy. We had prepared for years and years to make it possible.

The same is true of solar power satellites. The venture is huge and the path uncertain. That first step is the most difficult but will change the course of civilization for as far into the future as we can imagine. But just as our dream of sailing the oceans of the world started with one little incident and then built through the years to where it became reality, so has the path to space solar power been building through the years. Beginning with Peter Glaser's idea in 1968, it grew with the help of NASA engineers. It started to blossom with Department of Energy/NASA Systems Definition Studies in the late 1970's. It ignited the vision of a better world in the mind of a Boeing Engineer. Then it was staggered by the body blows of an administration and powerful nuclear foes in the Department of Energy and was nearly dead by the time it was rebuffed by the oil industry.

Through the years a few of us have

barely kept this idea alive while the technology has continued to evolved. The last great stumbling block of space transportation will be swept away when we are willing to make the investment.

● ● ● ● ● ● ● ● ● ● ● ●

As we step out into this great new adventure, the pathway will not always be smooth or clear. It is new territory that has never been trod before and politics will surely give it a bumpy ride. Public opinion will challenge and encourage. But if we move out courageously and with a clear goal in sight we will have the first satellite on line and there will be a stream of energy flowing from space. Like a dam bursting it will become a torrent. Other nations and peoples will want to join the new world of the 21st century.

We can do this. From the first energy flowing from space will come the demand for more and more. Japan, India and China are waiting to fill their needs with energy from space. Here in the United States the conversion from fossil fuels will be complete. The United States can become an energy exporter again and reap the benefits of a strong economy.

During this period of increasing energy from space the transportation industry will be moving towards electric cars. Our streets will be filled with the eerie quiet of electric cars. The congestion of our city streets will ease as we switch to the convenience of mass transient to commute to work or go into the city to shop. Gas stations will disappear. The sky will be clear again. We will eventually reach the point where we no longer need Middle East oil. Their economic leverage over the rest of the world will be gone and they will have to turn to a new source of income to survive.

Massive new industries will be created to manufacture the satellite hardware, transport it to space and assemble the satellites. On the earth the rectennas will need workers throughout the world to build and install them. Low-cost energy will become available to peoples that could not afford it before. There will be an opportunity for poor nations to evolve to a higher standard of living using low-cost non-polluting energy from space. One of the great advantages of a higher standard of living is improved education. Improved education has been shown to lead to reduced birthrates that could help stem the world's exploding population before it reaches the level where it can no longer be sustained.

At some point along this transition, oil consumption will drop to a level where it no longer is the dominate energy source. That will be the beginning of the fourth era, energy from space. Solar power satellites will ring the earth, delivering power to every nation on the earth. We will still be using other energy sources to supplement energy from space. Oil will no longer be king, it will still be available in limited quantities, but our world will not be dependent on it for commerce.

The level of carbon dioxide in the atmosphere will drop through natural processes as we will no longer be adding huge amounts through the burning of fossil fuels.

We will no longer be dependent on Middle East oil and we can leave them alone to work out their own problems without interference from the West. There will be no reason to fight wars over oil. Energy will be available to all who want and need it.

As we look into the night sky on a dark moonless night we may be able to see the solar power satellites as a faint string of pearls ringing the earth over the equator. They are like fountains pouring their life giving energy from the sky onto the earth. They will power the earth as far into the future as we can imagine. There is no end of the energy as long as the sun shines.

• • • • • • • • • • • •

Energy for the earth is not the only benefit that space solar power will bring. In order to build the satellites, the first thing that must happen is the development of a low-cost space transportation system to carry the hardware to space. With low-cost space transportation, space would be opened to all kinds of development. Industry will be able to move into space to process materials that benefit from microgravity. Manufacturing could take place there. Space hotels will become destination spots for adventurous vacationers. Imagine a week in space looking down on the panorama of the earth

flowing beneath you. Hotels shaped like as wheels that rotate and create partial gravity for your comfort, so you could enjoy having an elegant meal as you watched the sun set and then rise. This would just be the beginning.

With low-cost transportation to space, all of space would benefit all of mankind. Now those that have dreamed of going to the moon or Mars or beyond could realistically do that. It would no longer take a committed national program to go to the moon. Now it would be possible for individuals or companies to finance these kinds of ventures.

NASA would no longer need to use a great portion of their budget for transportation costs to accomplish their missions. They could simply contract to have one of the launch service companies drop them off in orbit for a small fraction of what it now costs. They could use their budget for much more advanced exploration of other parts of the universe.

Many adventurers on the earth yearn for colonization of space. This would now be possible. Permanent bases could be established on the moon. Mars has also held the fascination of many and has much greater potential than the moon. It even has a very thin atmosphere, and it seems there is some evidence of water. It will be a fascinating place to explore. The asteroids hold potential for finding minerals and raw materials that can supplement our supply here on the earth. As time goes by I expect we will venture on to the outer planets to check out their moons. Some of them may be a little like our earth. Seeing the rings of Saturn up close would be a spectacular sight.

The opening of space occurred in 1957 with the launching of Sputnik. What has been happening since can be compared to what happened after Columbus opened the new world in 1492. It took several centuries of development and colonization before the American colonies really started to blossom. Then with the independence of the United States in 1776, dramatic growth and expansion into the west created unparalleled expansion and wealth of mankind. That is what will happen in space. Since 1957 the development of space has evolved fairly rapidly over the ensuing decades, but the most exciting growth is yet to come. With the advent of solar power satellites, space will undergo a period of explosive development that will be even faster than the growth of the United States in the 19th and 20th century.

• • • • • • • • • • • • •

As we stand here today facing the problems of a global recession, war in the Middle East, the prospect of passing the peak of world oil production, and the devastating consequences of global warming, we have a decision to make. Do we continue to close our eyes to the problems and let our leaders blithely convince us not to worry, everything will be OK, or do we face up to reality and demand action. It is not that there isn't a solution; there is one as I have clearly shown you. The problem is that our most powerful industrial leaders and their lobbying efforts have been able to stop any progressive action that they perceive to be detrimental to their own interests. The country and the world be damned.

As we move into 2009 we are in a world economic crisis. However, our nation is taking the first giant steps towards a brighter future with a new generation of leadership in Washington DC. They stand on the threshold of history with the vision to lead the United States and the World into a new era of opportunity, peace, and care for the earth that will benefit all of mankind far into the future; the era of energy from space.

We must move forward with vision and courage. It is time to untie the lines and set out on our great new adventure.

• • • • • • • • • • • •

Latest Titles From Apogee Books

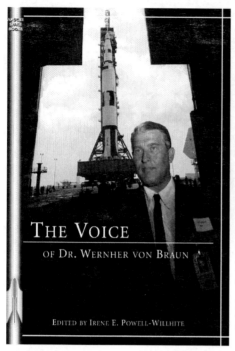

The Voice of Dr. Wernher von Braun

This book is a selection of the more than 500 speeches made by Dr. Wernher von Braun from 1947 to 1976. His passion as a crusader for worthy causes comes through clearly as he addresses education, the Cold War, religion, the space program, and more. von Braun is known primarily as the rocket scientist who made possible the Saturn V booster that put men on the Moon, but this book shows the true depth of the man as a human being. It is unique in that it presents von Braun's actual words, in context, and in contrast to the many books that have been written about him based on second-hand accounts.

ISBN 978-1-894959-64-3

Available NOW!

Astronautics Book 1

This first of two parts chronicles the initial discoveries, inventions, and engineering innovations that became the foundation of rocket technology. It follows the events which shaped the initial thrust into space, beginning with the first Soviet Sputniks and the shocked American response. The engineering requirements of the first manned spacecraft – the USSR's Vostok and US's Mercury – and the selection and experiences of the first spacefarers are all related in detail. Book 1 – *The Dawn of the Space Age* – tells the story up to the early 1970's. Space enthus-iasts will learn things that they never knew before from this definitive history of space exploration.

ISBN 978-1-894959-63-6

Available NOW!

Latest Titles From Apogee Books

Astronautics Book 2

Book 2 – *To the Moon and Towards the Future* – is the second of two parts examining the events leading America's commitment to land a man on the Moon in the 1960s and how that venture shaped the future of space exploration. It details the Gemini, Voskhod, Soyuz and Apollo programs and the exploration of the Moon. It also reviews the development of the Space Shuttle and details the evolution of the International Space Station. It high-lights the effort to find extraterrestrial life and the exploration of the outer planets. In addition, it examines advanced pro-pulsion technologies and speculates on what might lie ahead in space exploration.

ISBN 978-1-894959-66-7 Available NOW!

License to Orbit

Two leading experts on space systems with decades of experience in the field provide important and current insights into: - The Billionaire Players: The lives, ambitions and struggles of the billionaires who are funding and supporting the new commercial space industry. - The Global Perspective: Who are the global participants in the commercial space business and their strengths and weaknesses. - What is the nature of the strained relations and the alternative goals of space agencies versus private space flight industry. - Future Space Technologies: Descriptions of the amazing new technologies that could revolutionize space exploration and space industries in coming decades and how commercial systems might accelerate new trends. - Strategic Concerns with Private Space: What is the strategic nature of commercial space initiatives and the possible shifting US policy in this area - Interdisciplinary Perspective: The evolving interdisciplinary nature of commercial space in terms of launch technology, safety concerns, insurance, laws and regulation, financial requirements, and marketing strategies.

ISBN 978-1-894959-69-8 Available NOW!

Latest Titles From Apogee Books

ISScapades – The Crippling of America's Space Program

Donald A. Beattie, a former senior manager at the National Science Foundation, the Department of Energy and NASA, recounts the controversial evolution of the troubled International Space Station from the perspective of a participant and close observer who worked side by side with many of the early NASA Space Station managers. He pulls no punches in describing the political and managerial conflicts that resulted in severely compromising a major international program – a program that may never achieve the research goals for which it was envisioned.

ISBN 978-1-894959-59-9

Available NOW!

Krafft Ehricke's Extraterrestrial Imperative

This book provides an understanding of the early history of the space pioneers, what they helped to accomplish, and how Ehricke's vision of where we should be going can shape the future. At this difficult time, when there are questions about the future path of America's space program, Krafft Ehricke's vision-his Extraterrestial Imperative-lays out the philosophical framework for why space exploration must be pursued. Readers will find it an imaginative work, and an up-lifting story that contains a vast array of reasons why the human race needs to get off planet Earth and explore space. Krafft Ehricke's Extraterrestrial Imperative is the summation of a lifetime of work encouraging the exploration and development of space.

ISBN 978-1-894959-50-6   Available NOW!

Apogee Books Space Series

#	TITLE	ISBN	Bonus	US$	UK£	Can$
1	Apollo 8 NASA Mission Reports	978-1-896522-66-1	C	$18.95	£13.95	$25.95
2	Apollo 9 NASA Mission Reports	978-1-896522-51-7	C, w/o	$16.95	£12.95	$22.95
3	Friendship 7 NASA Mission Reports	978-1-896522-60-9	C	$18.95	£13.95	$25.95
4	Apollo 10 NASA Mission Reports	978-1-896522-68-5	C	$18.95	£13.95	$25.95
5	Apollo 11 NASA Mission Reports, Vol 1	978-1-896522-53-1	C	$18.95	£13.95	$25.95
6	Apollo 11 NASA Mission Reports, Vol 2	978-1-896522-49-4	C	$15.95	£10.95	$20.95
7	Apollo 12 NASA Mission Reports	978-1-896522-54-8	C	$18.95	£13.95	$25.95
8	Gemini 6 NASA Mission Reports	978-1-896522-61-6	C	$18.95	£13.95	$25.95
9	Apollo 13 NASA Mission Reports	978-1-896522-55-5	C	$18.95	£13.95	$25.95
10	Mars NASA Mission Reports	978-1-896522-62-3	C	$23.95	£18.95	$31.95
11	Apollo 7 NASA Mission Reports	978-1-896522-64-7	C, w/o	$18.95	£13.95	$25.95
12	The High Frontier	978-1-896522-67-8	C	$21.95	£17.95	$28.95
13	X-15 NASA Mission Reports	978-1-896522-65-4	C, w/o	$23.95	£18.95	$31.95
14	Apollo 14 NASA Mission Reports	978-1-896522-56-2	C	$18.95	£15.95	$25.95
15	Freedom 7 NASA Mission Reports	978-1-896522-80-7	C, w/o	$18.95	£15.95	$25.95
16	Shuttle STS 1-5 NASA Mission Reports	978-1-896522-69-2	C	$23.95	£18.95	$31.95
17	Rocket & Space Corporation Energia	978-1-896522-81-4		$21.95	£16.95	$28.95
18	Apollo 15 NASA Mission Reports	978-1-896522-57-9	C, w/o	$19.95	£15.95	$27.95
19	Arrows to the Moon	978-1-896522-83-8		$21.95	£17.95	$28.95
20	The Unbroken Chain	978-1-896522-84-5	C	$29.95	£24.95	$39.95
21	Gemini 7 NASA Mission Reports	978-1-896522-82-1	C	$19.95	£15.95	$26.95
22	Apollo 11 NASA Mission Reports, Vol. 3	978-1-896522-85-2	D, w/o	$27.95	£19.95	$37.95
23	Apollo 16 NASA Mission Reports	978-1-896522-58-6	C	$19.95	£15.95	$27.95
24	Creating Space	978-1-896522-86-9		$30.95	£24.95	$39.95
25	Women Astronauts	978-1-896522-87-6	C	$23.95	£18.95	$31.95
26	On To Mars	978-1-896522-90-6	C, w/o	$21.95	£16.95	$29.95
27	The Conquest of Space	978-1-896522-92-0		$23.95	£19.95	$32.95
28	Lost Spacecraft	978-1-896522-88-3		$30.95	£24.95	$39.95
29	Apollo 17 NASA Mission Reports	978-1-896522-59-3	C	$19.95	£15.95	$27.95
30	Virtual Apollo	978-1-896522-94-4		$24.95	£15.95	$27.95
31	Apollo EECOM	978-1-896522-96-8	C, w/o	$29.95	£23.95	$37.95
32	A Vision of Future Space Transportation	978-1-896522-93-7	C	$27.95	£21.95	$35.95
33	Space Trivia	978-1-896522-98-2		$19.95	£14.95	$26.95
34	Interstellar Spacecraft & Multi	978-1-896522-99-0		$24.95	£18.95	$30.95
35	Dyna-Soar	978-1-896522-95-1	D	$32.95	£23.95	$42.95
36	Rocket Team	978-1-894959-00-1	D	$34.95	£24.95	$44.95
37	Sigma 7 Mission Reports	978-1-894959-01-8	C	$19.95	£15.95	$27.95
38	Women of Space	978-1-894959-03-2	C	$22.95	£17.95	$30.95
39	Columbia Accident Report	978-1-894959-06-3	C, w/o	$25.95	£19.95	$33.95
40	Gemini 12 Mission Reports	978-1-894959-04-9	C	$19.95	£15.95	$27.95
41	Simple Universe	978-1-894959-11-7		$21.95	£16.95	$29.95
42	New Moon Rising	978-1-894959-12-4	D	$33.95	£23.95	$44.95
43	Moonrush	978-1-894959-10-0		$24.95	£17.95	$30.95
44	Mars Mission Reports, Vol 2	978-1-894959-05-6	D	$28.95	£20.95	$38.95
45	Rocket Science	978-1-894959-09-4		$20.95	£15.95	$28.95
46	How NASA Learned to Fly	978-1-894959-07-0		$25.95	£18.95	$35.95
47	Virtual LM	978-1-894959-14-8	C	$29.95	£22.95	$42.95
48	Deep Space Mission Reports	978-1-894959-15-5	D	$34.95	£22.95	$44.95

Bonus legend: C = CD-ROM, D = DVD, w/o = available from our web site only.

http://www.apogeebooks.com

Apogee Books Space Series

#	TITLE	ISBN	Bonus	US$	UK£	Can$
49	Space Tourism	978-1-894959-08-7		$20.95	£15.95	$28.95
50	Apollo 12 Mission Reports, Vol 2	978-1-894959-16-2	D	$24.95	£15.95	$31.95
51	Atlas	978-1-894959-18-6		$29.95	£16.95	$37.95
52	Reflections from Orbit	978-1-894959-22-3		$23.95	£16.95	$30.95
53	Real Space Cowboys	978-1-894959-21-6	D	$29.95	£17.95	$36.95
54	Saturn	978-1-894959-19-3	D	$27.95	£18.95	$35.95
55	On To Mars 2	978-1-894959-30-8	C	$22.95	£14.95	$29.95
56	Getting Off the Planet	978-1-894959-20-9		$18.95	£12.95	$23.95
57	Return to the Moon	978-1-894959-32-2		$22.95	£15.95	$28.95
58	Beyond Earth	978-1-894959-41-4		$27.95	£18.95	$36.95
59	ISScapades	978-1-894959-59-9		$23.95	£12.95	$27.95
60	Astronautics Book 1	978-1-894959-63-6		$23.95	£12.95	$27.95
61	Go For Launch	978-1-894959-43-8		$29.95	£18.95	$34.95
62	Reference Guide to ISS	978-1-894959-34-6		$21.95	£10.95	$23.95
63	Around World 84 Days	978-1-894959-40-7		$31.95		
64	Spaceships	978-1-894959-50-6		$19.95	£10.95	$22.95
65	Voice of Wernher von Braun	978-1-894959-64-3		$22.95	£12.95	$24.95
66	Surveyor Mission Reports	978-1-894959-65-0		$17.95	£9.95	$21.95
67	Astronautics Book 2	978-1-894959-66-7		$25.95	£13.95	$26.95
68	End of the Solar System	978-1-894959-68-1		$26.95		
69	Lunar Exploration Scrapbook	978-1-894959-69-8		$36.95	£16.95	$36.95
70	Apollo Advanced Lunar Exploration Plan	978-1-894959-80-3	w/o	$16.96		
71	Canada's Fifty Years in Space	978-1-894959-72-8		$26.95		
72	Floating to Space	978-1-894959-73-5	D	$27.95		
73	The Astronaut & The Fireman	978-1-894959-74-2		$TBA		
74	Saturn I/IB	978-1-894959-85-8	D	$26.95		
75	Cold War Tech War	978-1-894959-77-3		$29.95		
76	Krafft Ehricke's Extraterrestrial Imperative	978-1-894959-91-9		$27.95		
77	License to Orbit	978-1-894959-98-8		$27.95		
78	The Nuclear Rocket	978-1-894959-99-5		$21.95		
79	Apollo 17 NASA Mission Reports Vol 2	978-1-926592-02-2	D	$26.95		
80	Lightcraft Flight Handbook	978-1-926592-03-9		$29.95		
81	Energy Crisis: Solution From Space	978-1-926592-06-0		$24.95		
82	Rocket Belt Pilot's Manual	978-1-926592-05-3		$TBA		
83	Selling Peace	978-1-926592-08-4		$TBA		
84	The Farthest Shore	978-1-926592-07-7		$TBA		

Apogee Books Pocket Space Guides

#	TITLE	ISBN	US$	UK£	Can$
1	Apollo 11	978-1-894959-27-8	$9.95	£6.95	$12.95
2	Mars	978-1-894959-26-1	$9.95	£6.95	$12.95
3	Deep Space	978-1-894959-29-2	$9.95	£6.95	$12.95
4	Launch Vehicles	978-1-894959-28-5	$9.95	£6.95	$12.95
5	Apollo Test Program	978-1-894959-36-0	$9.95	£6.95	$10.95
6	Apollo Exploring	978-1-894959-37-7	$9.95	£6.95	$10.95
7	Hubble Telescope	978-1-894959-38-4	$9.95	£6.95	$10.95
8	Russian Spacecraft	978-1-894959-39-1	$9.95	£6.95	$10.95
9	Project Constellation	978-1-894959-49-0	$9.95	£6.95	$11.95
10	Space Shuttle	978-1-894959-52-0	$9.95	£6.95	$11.95
11	Project Mercury	978-1-894959-53-7	$9.95	£6.95	$11.95
12	Project Gemini	978-1-894959-54-4	$9.95	£6.95	$11.95